Unintended Outcomes of Social Movements

Why did the 1989 Chinese student movement end in violent confrontation at Tiananmen Square, despite the fact that both the Chinese government and the students very much wanted to avoid violence? This puzzle, which lies at the heart of the tragic events at Tiananmen, is addressed here from a fresh perspective that sheds new light on these dramatic events.

Throughout *Unintended Outcomes of Social Movements*, Deng applies the formal methods of game theory to elucidate some of the contingent, strategic decision making by both sides in a social-movement/state confrontation, and how those decisions can—and did—lead to an unintended outcome. In identifying the underlying cause of the Tiananmen tragedy, namely a newly created social system with four highly specific properties, this book provides the first adequate explanation of the Tiananmen events. It therefore stands to make a significant stride toward convincing students of political conflict of the explanatory power of formal game theoretic models.

This book is an excellent source of reference for both undergraduate and postgraduate students in areas including Chinese politics, social movements, game theory economics, and social theory.

Fang Deng is Associate Professor in the department of Sociology at Bridgewater State College, US. Previous publications include Chinese translations of *Game Theory and Economic Modeling* by David M. Kreps (Oxford University Press, 1992) and *Foundations of Social Theory* by James S. Coleman (Harvard University Press, 1990).

International Library of Sociology
Founded by Karl Mannheim
Editor: John Urry, Lancaster University, UK

Recent publications in this series include:

Risk and Technological Culture
Towards a sociology of virulence
Joost Van Loon

Reconnecting Culture, Technology and Nature
Mike Michael

Advertising Myths
The strange half lives of images and commodities
Anne M. Cronin

Adorno on Popular Culture
Robert R. Witkin

Consuming the Caribbean
From Arawaks to Zombies
Mimi Sheller

Between Sex and Power
Family in the world, 1900–2000
Goran Therborn

States of Knowledge
The co-production of social science and social order
Sheila Jasanoff

After Method
Mess in social science research
John Law

Brands
Logos of the global economy
Celia Lury

The Culture of Exception
Sociology facing the camp
Bülent Diken and Carsten Bagge Laustsen

Visual Worlds
John Hall, Blake Stimson, and Lisa Tamiris Becker

Time, Innovation and Mobilities
Travel in technological cultures
Peter Frank Peters

Complexity and Social Movements
Multitudes acting at the edge of chaos
Ian Welsh and Graeme Chesters

Qualitative Complexity
Ecology, cognitive processes and the re-emergence of structures in post-humanist social theory
Chris Jenks and John Smith

Theories of the Information Society, 3rd Edition
Frank Webster

Crime and Punishment in Contemporary Culture
Claire Grant

Mediating Nature
Nils Lindahl Elliot

Haunting the Knowledge Economy
Jane Kenway, Elizabeth Bullen, Johannah Fahey, and Simon Robb

Global Nomads
Techno and New Age as transnational countercultures in Ibiza and Goa
Anthony D'Andrea

The Cinematic Tourist
Explorations in globalization, culture and resistance
Rodanthi Tzanelli

Non-Representational Theory
Space, politics, affect
Nigel Thrift

Urban Fears and Global Terrors
Citizenship, multicultures and belongings after 7/7
Victor J. Seidler

Sociology through the Projector
Bülent Diken and Carsten Bagge Laustsen

Multicultural Horizons
Diversity and the limits of the civil nation
Anne-Marie Fortier

Sound Moves
IPod culture and urban experience
Michael Bull

Jean Baudrillard
Fatal theories
David B. Clarke, Marcus A. Doel, William Merrin, and Richard G. Smith

Aeromobilities
Theory and method
Saulo Cwerner, Sven Kesselring, and John Urry

Social Transnationalism
Steffen Mau

Towards Relational Sociology
Nick Crossley

Unintended Outcomes of Social Movements
The 1989 Chinese student movement
Fang Deng

Forthcoming in the series:

Global China
Lash Scott, Keith Michael, Arnoldi Jakob, and Rooker Tyler

Stillness in a Mobile World
David Bissell and Gillian Fuller

Revolt, Revolution, Critique
The paradox of society
Bulent Diken

Unintended Outcomes of Social Movements

The 1989 Chinese student movement

Fang Deng

Routledge
Taylor & Francis Group
LONDON AND NEW YORK

First published 2011
by Routledge
2 Park Square, Milton Park, Abingdon, Oxon OX14 4RN

Simultaneously published in the USA and Canada
by Routledge
270 Madison Avenue, New York, NY 10016

Routledge is an imprint of the Taylor & Francis Group

Typeset in 10/13pt Sabon by
Book Now Ltd, London
Printed and bound in Great Britain by
CPI Antony Rowe, Chippenham, Wiltshire

British Library Cataloguing in Publication Data
A catalogue record for this book is available from the British Library

Library of Congress Cataloging in Publication Data
Deng, Fang, 1949–
Unintended outcomes of social movements: the 1989 Chinese student movement/by Fang Deng.
p. cm.
Includes bibliographical references and index.
1. China—History—Tiananmen Square Incident, 1989 2. Student movements—China. 3. Social movements—China. 4. Game theory.
I. Title.
DS779.32.D46 2010
951.05'8—dc22 2010010436

ISBN: 978–0–415–77933–3 (hbk)
ISBN: 978–0–203–86889–8 (ebk)

Contents

Illustrations

Tables

Figures

Preface

"Are we going where we want to go?" This is a question people may ask when they drive, and it is also a question that is often asked when people build societies and become involved in creating norms and social values, initiating status distributions, and dealing with social conflict, etc. Unintended outcomes or unexpected results occur when people are lost. Nowadays, digital technology has provided cars with GPS to guide us in getting where we want to go; but how do we choose the right way to go in constructing a better society? There is demand for a social GPS, and sociologists may be able to provide it. Social navigation will necessarily consist of many parts, and this book is intended to provide one of them, which relates to dealing with social conflict.

The distinguishing feature of this book is the fact that it studies a social movement, but emphasizes the importance and consequences of the newly created social system that emerges when social movements are formed. This new social system is composed of a movement and a countermovement; in the case of the Tiananmen tragedy, the countermovement was the Chinese government. The development of the 1989 Chinese student movement initiated a process in which the movement and the government interacted and generated a new social system with four intrinsic properties: movement participants' anti-threat resistance, state's sub-optimal strategies, short-term gain and long-term loss for movement participants, and the development of an information gap between the movement and the state. This newly created social system, with its four intrinsic properties, was a necessary and sufficient condition to produce the tragic events that unfolded at Tiananmen. A clearer understanding of the ways in which this newly created social system works can lead to better outcomes for both movement participants and governments.

There could be no in-car GPS without digital technology. Likewise, a deeper understanding of social systems cannot be constructed without adequate research methods. This book therefore pursues methodological approaches that go beyond the routine modes of explanation. By applying principles of game theory to the study of the 1989 Chinese student movement, the analysis of equilibriums in the theory of non-cooperative games is demonstrated to be a powerful tool in analyzing interactions between the strategic choices made by a movement and those made by a state. More specifically, techniques from game theory permit incorporation of the element of time and the availability of information—and their important impacts—into the study of social movements and their outcomes. This approach therefore provides an alternative to existing research methods and may lead to a better understanding of social systems.

Acknowledgments

I would like to thank the many people to whom I am indebted for their valuable help and comments. I was uncertain about this project in the beginning, but James S. Coleman encouraged me to continue my work on it. James was widely regarded as an outstanding sociologist and, without his vision, this book could not have been conceived. In addition, William L. Parish has been very enthusiastic about my project, and gave me the opportunity to defend my work before hostile workshop audiences. Michael Chwe and Roger Gould gave me technical advice on game theoretic analysis. Without their commitment and strong support this book would not exist. The reviewers for *American Journal of Sociology* provided me with detailed and perceptive comments on the paper that forms the basis for this book. Their careful assistance was particularly helpful and much appreciated. I am also grateful to Corinna Fales, who edited the book with skill and grace. Finally, I am deeply grateful to John Urry, the editor of the International Library of Sociology, and Gerhard Boomgaarden, publisher of Routledge, for their confidence in the book.

1 Understanding unintended outcomes of social movements

Two photographs taken in 1989 have repeatedly appeared all over the world. In one, thousands of Germans stand on the Berlin Wall celebrating German reunification; in the other, a lone Chinese student stands in front of a military tank at Tiananmen Square and tries to block the other tanks that surround him. What very different pictures!

In the late 1980s and early 1990s, while the world cheered the collapse of communism in the former Soviet Union and other East European countries, it condemned the Chinese government for its suppression of the Tiananmen protests. Since then, these two photographs have been used to symbolize a typology of states based on the concept of democracy: democratic vs. non-democratic regimes. Although the categories included in the typology are very simple, reality in China is much more complex.

While the famous photo of the lone Chinese student at Tiananmen was used with another photo of the Berlin Wall to explain the historical events of 1989, many features and developments within the Chinese student movement have been neglected in the literature, and the public is not knowledgeable about them. Although more than 20 years have passed, it is necessary to know what really happened during the 1989 Chinese student movement in order to comprehend many of the significant changes currently taking place in China. As Snow, Soule, and Kriesi argue:

> Social movements and the activities they sponsor have become a kind of fifth estate in the world today. If so, then understanding our own societies, as well as the larger social world in which they are embedded, clearly requires some knowledge and understanding of social movements and the activities with which they are associated.
>
> (2004:3)

In *Unintended Outcomes of Social Movements*, I go beyond that snapshot of the student at Tiananmen Square and unfold both the vivid features of the movement and some implications for policy.

An unintended outcome of the 1989 Chinese student movement

Here, a critical question arises. Was the June massacre of 1989 in Beijing an intended or an unintended outcome for the movement and the state? If it was an intended outcome for both sides, why did the Chinese government quell the people, and why were movement participants not afraid of being killed? If it was an intended outcome for only one side, what prevented the other side from avoiding the tragedy? And if it was not the intended outcome for both sides, why was the movement suppressed by violence? It is very important to answer these questions because tragedies like Tiananmen cannot be prevented from occurring again without understanding how and why they unfolded as they did.

In order to fully understand the questions and arrive at the correct answers, it is necessary to look at the state and the movement side by side but separately. First, let us consider the movement. Was the June massacre of 1989 in Beijing an intended or an unintended outcome for the movement? One of the most significant aspects of the Chinese student protests was that the massacre was a totally unexpected outcome for the movement. The majority of its participants were students from China's most prestigious universities who were poised to enjoy unprecedented opportunities and bright futures. While the whole world watched these brave students fight for democracy in China, the students themselves did not expect their activism to present any danger, either before or after the declaration of martial law.

Some foreign eyewitnesses, for instance, experienced the movement's spirit in May 1989. According to *China Daily* (an English newspaper published in China), foreign business people expressed both surprise and concern when more than 1,000,000 people poured onto the streets in support of student hunger strikers. The Deputy General Manager of NTT (Nippon Telegraph and Telephone Corporation) Beijing's Representative Office told *China Daily* on 17 May that he and his colleagues were initially very nervous about the demonstrations; but, when they found many marchers in a jubilant spirit, their fears began to ease (*China Daily*, 18 May 1989).

In his 1992 essay, Tang Tsou explained why the students were unwilling to withdraw from Tiananmen Square after the declaration of martial law:

> At that time, there was a widespread belief that the People's Liberation Army would not turn its guns on the people. It is impossible to know who first began promoting that opinion and how it became so widely accepted. But it is true that many students and Beijing residents believed that the army was "the people's army" and would not carry out any orders to harm the people.
>
> (1992:313)

Timothy Brook described Beijing's great shock at the massacre: "It was not only the gruesome spectacle of violent death that shocked. It was also the shock of surprise" (1998:5). "No one was more surprised by the live fire than the people themselves" (1998:10). Brook also provided some reasons for the Chinese people's reaction:

> A decade of economic reform and slow political liberalization through the 1980s induced them to believe that the harsh politics of the 1950s and 1960s—the liquidation of the landlords, the Hundred Flowers campaign, the anti-rightist purges, the Great Leap Forward, the Cultural Revolution—were nothing but faded nightmares from the past. And anyone in China could have told you that none of these campaigns was brought to conclusion by ordering soldiers to fire in the streets. Once, during the Cultural Revolution, a local commander in Guizhou province ordered his troops to fire on Red Guards, but it happened only once, without central authority, and the commander was disciplined for his action. Who in 1989 could imagine that violence would be brought to bear this time?
>
> (1998:5)

In the afternoon of 3 June 1989, several hours before the massacre began, the Joint Federation of All Circles in Beijing (the top leadership of the movement) held a meeting to discuss the situation. Although most leaders realized that it was getting serious, none believed that the government would use force to suppress the movement.

> As late as the early hours of Sunday morning, recalls a Beishida graduate, he and his friends "figured the worst that could come of

> this would be a wave of arrests, aimed particularly at the student leaders." They could not imagine slaughter. Another Beishida resident presumed the same thing that night. "We knew that something was going to happen. We thought they would move in, use tear gas, and rough them up a little—or a lot—and throw them in jail. We never expected that they would go in there shooting."
>
> (Brook, 1998:197)

At about 10:30 p.m. on 3 June 1989, the military crackdown began at Muxudi, an intersection on West Changan Boulevard, a few miles away from Tiananmen Square. At the same time, the opening ceremony of a new university—the Democracy University, which was established by the students themselves—was in progress at the square. The large crowd was very enthusiastic. There were more than 100,000 people at the square, and most of them sensed that the government might be planning to use some new methods to enforce martial law; but they assumed that, no matter what methods were used, none of them would work because nothing dire had happened since martial law was implemented on 20 May. However, while people celebrated the opening of the Democracy University at Tiananmen Square, the military crackdown had begun few miles away.

Hundreds of thousands of people, including pregnant women and teenage girls, gathered to block the troops at Muxudi. When gunfire began, people there could not even grasp that a military crackdown was finally taking place. Most of them had thought that the soldiers were simply driving people off the streets by using rubber bullets. Almost all of the innocent students and Beijing residents who were killed had believed that the People's Army would never hurt them.

In short, all studies and evidence point in one direction—namely, that the June massacre of 1989 in Beijing was an unexpected and unintended outcome for the movement. But what was the other side's story? Was the massacre an intended or an unintended outcome for the Chinese government?

Based on his study of a series of dialogues between high-ranking officials and student representatives in May 1989, Tang Tsou believed that the massacre was not the intended outcome for the Chinese government:

> By having asked two top officials, one in the party and the other in the government, to conduct a dialogue with the students, the regime had given de facto recognition to the students without

> explicitly renouncing the April 26 editorial. There was at least no attempt to suppress them.
>
> (1992:306)

The April 26 editorial was the lead editorial in the Chinese Communist Party's organ, *People's Daily*. It characterized the student movement as a "planned conspiracy" and "turmoil."

In contrast to Tang Tsou, Timothy Brook believed that for the Chinese government "the Massacre was a desired outcome" (Brook, 1998:196). He provided three forms of evidence for his conclusion, which are challenged below.

First, Brook drew an inference from Deng's public speech:

> When Deng Xiaoping appeared on television on June 9th to praise his martial law commanders, he insisted that the suppression had altered nothing fundamental in the Chinese government's course. . . . Deng was not disconcerted by the destruction or the response around the world. "The storm was bound to happen sooner or later," he declared. "It was just a matter of time and scale." His words project an image of supreme confidence, of a leader who enjoys full control of his government and his military. They also suggest that Deng regarded both the Democracy Movement and its suppression as inevitable. . . . Deng may have been doing nothing more than wrapping himself in the mantle of Communist wisdom. Even so, the remark implies that the Massacre was a desired outcome. Deng would have us believe that the operation went as planned; he offered no sign of contrition or regret.
>
> (1998:196)

Although it is true that Deng Xiaoping had made it clear that everything went smoothly, as planned, evidence that an outcome is intended or unintended, which is based solely upon the speeches of top governmental officials, is not a reliable source. This is because, as is common knowledge in China, "it is impossible for the government to admit its mistake openly" (Tsou, 1992:306). If Deng Xiaoping had publicly implied that the massacre was not a desired outcome, it would have meant that he, as sole decision maker, had made a serious strategic mistake. In China, no one expected either Deng Xiaoping or any other top government official to admit their mistakes openly or to offer any sign of contrition or regret after the massacre.

Brook's second piece of evidence was the military preparation:

> Certainly, the scale of the operation, involving at the very least 150,000 soldiers along with substantial armored support, testifies to considerable planning. The first order bringing outside troops into the Beijing Military District was issued before April 25th. Military leaves were cancelled by the beginning of May. At least two-thirds of the final force was in position for the first assault on May 19th. From the first, Deng considered the army an appropriate means for stopping the students' protests. He sided with it to the last.
>
> (1998:197)

There is no doubt that the Chinese government prepared the military to halt the student movement. But there was a delay of 38 days in the military crackdown between 25 April and 3 June. Why? What does this delay tell us about the government's intention? Even Brook believed that the Chinese government intended to stop the movement in a non-military way, which represents an inconsistency in his opinions. He points out that: "The delay in bringing military means to bear on the movement does suggest a willingness to try other means first, though always with the understanding that, if a political solution failed, a military solution would follow" (1998:197).

During the delay, the Chinese government repeatedly threatened students, and then backed down when they resisted. On 26 April 1989, for example, when Chinese officials sent a signal through an editorial in the Communist Party organ, *People's Daily*, that the student protests would no longer be tolerated, they used the word "turmoil" to justify any crackdown they might choose to make. But, on 27 April, when students marched in Beijing, thousands of armed police officers withdrew from their position in front of the students and more than 200,000 bystanders. And, after 27 April, the Chinese government accepted the movement's demand to arrange dialogues with student representatives and Beijing journalists on seven separate occasions before declaring martial law. These facts demonstrate that the Chinese government intended to try to reach a political solution and avoid a bloody confrontation.

Likewise, on 20 May, when the government declared martial law and deployed thousands of troops, accompanied by tanks, and 1,000,000 people protested against the implementation of martial law, the troops and tanks were called off. Beijing residents spontaneously gathered at all the major entrances to the city, forming barricades and building roadblocks in an attempt to prevent military troops from entering the city. For 14 days, from 20 May to 2 June 1989, troops, trucks, and tanks were waiting at Beijing's outskirts for orders to move. During these 14 days, there were no clashes between military troops and the students. The government again demonstrated its intention to avoid a bloody confrontation. Despite heat and hunger, soldiers had to remain in the army trucks, and Beijing residents brought them bread, porridge, salted vegetables, and water. On 23 May, some of the troops began to retreat, and it appeared that the threat of military suppression was over.

At a meeting on 27 May, organized by the Joint Federation of All Circles in Beijing, all leaders voted to hold a large-scale demonstration to celebrate victory of the student movement, followed by withdrawal from Tiananmen Square on 30 May. It seemed that the movement would end peacefully. Unfortunately, student leaders from the General Headquarters of Tiananmen Square soon changed their minds (discussed in Chapter 5). But all records show that the Chinese government made every effort to avoid bloodshed while it prepared the military force to stop the student movement.

Finally, Brook provides evidence of an intended outcome for the Chinese government based on the action itself: military crackdown.

> The ferocity of the assault signals that, once the order to fire guns was given, this operation was designed—or, at the very least, expected—to cause violence and produce civilian casualties. No government sends tens of thousands of combat troops into a volatile urban setting without expecting violence to erupt. Pointing AK-47s at unarmed civilians is a formula for disaster.
>
> (1998:199)

Intention and action are different; different intentions may produce the same action. Therefore, it is inadequate to equate an action with an intention without a concrete analysis of the context of the action.

On the afternoon of 3 June 1989, several hours before the military crackdown occurred, the headquarters of the martial-law troops and the municipal government of Beijing had issued some public notices and this announcement: "The situation in Beijing at present is very serious. More serious riots can occur at any time. From now on, please do not come onto the street and do not go to Tiananmen Square" (*China Daily*, 4 June 1989). At 6:30 p.m., a few hours before the military crackdown occurred, all government-controlled radio and television had begun broadcasting this announcement repeatedly, throughout the city and the evening, warning people not to go out on the streets or "they should bear responsibility for their own fates" (*China Daily*, 4 June 1989). It was obvious that the government officials were trying to minimize casualties when high-ranking leaders finally decided to end the movement by military crackdown.

On 3 June, the massacre began at about 10:30 p.m. at Muxudi, a few miles away from Tiananmen Square. There are many indications that the government did not anticipate the consequences of the military crackdown on that evening.

> One foreign reporter thought to judge the degree of premeditation by seeing how the television news at 8:00 a.m. Sunday morning handled the incident. "The announcement was ready for broadcast, yet it was put together without pictures. The announcers were wearing all black (a gesture that would be forbidden later that day). They looked like they were going to cry. We saw their faces only briefly, and then it became voice-over. I can't understand why they didn't have any visual footage. If they had planned the massacre, they would have had the whole thing ready and used military announcers. The first news is important, and they screwed it up." The lack of preparedness in hospitals points to a similar failure to anticipate the scale of violence that night.
>
> (Brook, 1998:198–99)

Troops finally arrived at Tiananmen Square at about 1:30 a.m. on 4 June and began to encircle the square, where thousands of students remained. At 3:30 a.m., two intellectuals served as representatives of the movement and negotiated with the army for students' peaceful withdrawal from the square. The army accepted their demand, which shows that the government intended to avoid further bloodshed.

In sum, although the world has not forgotten the students who gave their lives for the dream of democracy in China, and has not stopped condemning the Chinese government for the massacre, the record clearly shows that the Chinese government tried very hard to avoid violence during the movement's development, and that the June massacre of 1989 was an unintended outcome for the government as well as for the student movement. The government, in fact, had a powerful reason to strive for a political solution instead of violence.

In 1989, Deng Xiaoping, the supreme leader of the Chinese government, was fully aware that three different outcomes of the Chinese student movement were possible. The first one was stopping the movement in a peaceful and non-military way. This would be the best outcome for the state because the one-party dictatorship would remain intact and Deng's economic reforms could continue without interruption.

Since 1978, Deng Xiaoping has pushed for a series of reforms and openness that can be called Deng's "second revolution," following the first one led by his predecessor, Mao Tse-tung. In 2005, George J. Church described the core principle of Deng's revolution:

> In the Chinese spirit of balance between yin and yang, Deng's second revolution is an attempt on a monumental scale to blend seemingly irreconcilable elements: state ownership and private property, central planning and competitive markets, political dictatorship and limited economic and cultural freedom. Indeed, it is almost, or so it often seems to skeptics in both the Western and Marxist worlds, an attempt to combine Communism and capitalism.
>
> (2005:1)

In 1985, Deng was *Time* Magazine's "Man of the Year." He had changed the daily lives of China's citizens to a greater extent than any other world leader had changed the lives of their constituents. Shortly after returning to power in 1978, Deng faced a fundamental problem that existed in all the communist countries:

> Since the October Revolution [of 1917, which enthroned Soviet Marxism), more than 60 years have passed. How is it that many socialist countries have not been able to overtake capitalist ones in terms of development? What was it that did not work? . . . Before Deng, the failure was more starkly obvious in China. The average peasant or city worker was little better off, if at all, when Mao died in 1976 than he or she had been in the 1950s.
>
> (Church, 2005:3)

Deng was determined to fix the problem by starting his "second revolution" but was criticized by his opponents within the party for betraying Marxist principles. The 1980s were a critical period for implementation of Deng's policies of reforms and openness. In 1989, although Deng presented fresh evidence of economic development—GDP rose from

354.82 billion RMB (Chinese dollars) in 1978 to 1,068.24 billion RMB in 1989 at constant prices (*China Statistical Yearbook*, 2009:39)—he desperately wanted to keep China as stable as possible in order to be able to continue to import foreign capital, technology, management methods, etc. Steel-hard Deng was determined to push his reforms through to their conclusion, whatever that might be, and wanted a political solution to the student demonstrations in order to minimize any negative impacts on his economic programs.

The second possible outcome was to stop the student movement by using military force. Deng and the other government leaders knew that a military crackdown on the nascent movement for democracy would almost certainly be followed by condemnation and serious economic sanctions from the international community. At the time, no one knew exactly how such measures would hurt China's economy. But Deng's policies of reforms and openness were still fragile, and any interruption would have put his second revolution at serious risk. Therefore, the Chinese government made every effort to avoid bloodshed.

The last possible outcome was that the Chinese Communist Party would collapse and the state would lose all of its power. This would be the worst possible outcome, because, for Deng Xiaoping, the core of his second revolution was the retention of Communist Party dictatorship in China. As a major economic policymaker, Deng was a pragmatist. "His most celebrated saying is a homely metaphor to the effect that it does not matter whether a cat is black or white so long as it catches mice" (Church, 2005:4). But Deng had no intention of letting China's political system morph into anything else. Although the communist parties in the former Soviet Union and East European countries had collapsed, Deng Xiaoping and his colleagues would never be willing to accept this fate as an outcome of the student movement.

In sum, there is no doubt that the June massacre of 1989 in Beijing was an unintended outcome not only for the movement but also for the Chinese government. This puzzle, then, lies at the center of the 1989 Chinese student movement: Why did it result in violent suppression, at Tiananmen Square, despite the fact that both the Chinese state and the activists wanted very much to avoid violent confrontation?

Like all social scientists, sociologists play two roles. They are scientists when they try to explain the world, and they are policymakers when they try to improve the world. Sociologists' responsibility as scientists and policymakers makes the study of the unintended outcome of the Chinese student movement, and the puzzle lying at its center, extremely important.

As policymakers, sociologists must satisfy a public demand for knowledge about unintended outcomes of social movements. Suppose another student movement emerges and the state again comprises the sole countermovement. Both the student organizations and state officials would like to rely on the advice of sociologists in order to help them choose the right strategies and reach their intended goals. If one of their goals is to avoid bloodshed, they may ask what the root causes were of the 1989 Tiananmen tragedy, under what conditions unintended outcomes of social movements occur, and what can be done to prevent such unintended outcomes from happening in the future.

As advisers and policymakers, sociologists must know the answers to the questions. Studying and understanding the unintended outcome of the 1989 Chinese student movement can significantly impact policymaking.

As scientists, sociologists must also satisfy a professional demand to develop a theory of unintended outcomes of social movements in which the unintended outcome is a balance or equilibrium at the system level. The emergence of a social movement generates a new social system composed of a movement and a countermovement. (In the case of the Tiananmen tragedy, the countermovement was the Chinese government.) This newly created social system has its own properties, on which the movement and countermovement base their actions. When social conflict—an unintended outcome of social movement—occurs at the macro, or system, level, James S. Coleman believes that:

> The task of the social theorists, as I came to see it, was to discover in real social systems implicit rules and norms, constraints and goals, and the way in which the actions they generate combine and interact to produce system functioning.
>
> (1994:33)

When sociologists try to explain the events at Tiananmen, it becomes necessary to undertake the study of unintended outcomes.

Two reasons why the Tiananmen tragedy has not been presented as an unintended outcome

Numerous works on the 1989 Tiananmen tragedy have been published, but none has presented the event as an unintended outcome of the student movement. As a result, one of the most important aspects of the movement—the fact that the violence was an unintended outcome—has been totally neglected. This neglect makes existing explanations of the Tiananmen tragedy incomplete or even misleading. If it is so important to study unintended outcomes, why has no one presented the events at Tiananmen Square in this way?

There are two fundamental reasons for this. First, the typology of states based on the concept of democracy has lagged behind the times. "What counts in evaluating a typology is not whether it is 'true' or 'false', but whether it is useful, and for what purpose" (Kesselman, Krieger, and Joseph, 2003:16). It is not possible to present the 1989 Tiananmen tragedy as an unintended outcome based on this typology, nor is the typology adequate for developing a theory to explain the tragic events.

For people who still use the typology of states based on the concept of democracy, it seems politically incorrect to think that Chinese Communist Party leaders did not want a military crackdown. Given China's undemocratic political system, these people believe that the violence at Tiananmen was inevitable.

For example, Timothy Brook thought that, "The Chinese Communist Party had to suppress the movement if it wished to continue its monopoly of state power" (1998:208). He described China's political system: "After winning power in 1949, the Communist Party did not sponsor the creation of a new state, but became it. On the people's behalf—and for their own good—the party monopolized all political power and exercised totalistic control" (1998:195). For Timothy Brook, "It [the Beijing massacre] occurred for the specific reason that the party as a state had available to it military means and chose to direct those means against the civilian population of Beijing" (1998:196).

In his 2001 study of the causes and development of the 1989 movement, Dingxin Zhao depicted different social movement dynamics under democratic and authoritarian regimes:

> A society tends to be radical in an authoritarian state and conformist in a democratic state. This proposition depicts the distinctive nature of state–society relations under different types of state. It implies that opposition movements tend to be radical in authoritarian states and reformist in democratic states—a crucial difference of social movement dynamics.
>
> (2001:27)

Although Tang Tsou (1992) believed that the massacre was not the intended outcome for the Chinese government, he explained the Tiananmen tragedy from a historical perspective, rather than addressing what can be done to prevent unintended outcomes of social movements.

> Both the extremist students and the old comrades acted according to the tradition of twentieth-century Chinese politics and philosophy: Politics is ultimately a life-and-death struggle of revolution against "autocracy," "dictatorship," "feudalism," or their remnant reactionary forces, or else it is revolution against counter-revolution. In the confrontation that sooner or later brings about an outcome in which one side wins and retains all power, and the other side loses it, no measure, however brutal and immoral, or however romantically unrealistic, is excluded, not even by the inevitability of long-term adverse consequences for the nation.
>
> (1992:294)

Based on this belief, Tsou specifically examined two key factors.

> That pattern of disagreement, conflict, struggle, confrontation, and outcome—as well as the rules and norms of a political game played under a structure of power with a one-man core—must be borne closely in mind in order for us to understand the Tiananmen tragedy, particularly the alternatives available to Zhao, his choices, and the final outcome.
>
> (1992:293)

Studies of the 1989 student movement have one thing in common: a focus on China's political system and how events at Tiananmen were, to a certain extent, determined by the undemocratic regime. Brook's monopoly of state power, Zhao's authoritarian regime, and Tsou's structure of power with a one-man core have all characterized China's political system. While none of these studies has presented the Tiananmen tragedy as an unintended outcome, they are politically accurate in that the military crackdown did occur in an undemocratic and communist country. They also help readers to understand the political context in which the tragedy occurred.

Unfortunately, however, these studies share a common weakness. They do not provide any clarification about the conditions under which the Tiananmen events occurred. In other words, China's political regime is neither a sufficient nor a necessary condition for the Tiananmen tragedy, and therefore does not explain it.

A condition is considered sufficient if it has existed and then the event occurs. In the 1980s, there were many student movements in China and all of them ended peacefully except that of 1989. If China's political regime were a sufficient condition for the military crackdown, all of these student movements would have ended in bloodshed. When scholars explain the Tiananmen tragedy by citing China's political system, they fail to address one simple but critical question: Why were the dynamics and outcome of the 1989 movement totally different from those of other student movements, when all of them occurred under the same non-democratic political regime?

Moreover, a condition is considered necessary for an event if the condition does not exist and then the event does not occur. But the event—a violent confrontation between a social movement and a countermovement—has, in fact, occurred in countries where a democratic political regime does exist. For example, on 4 May 1970, members of the Ohio National Guard fired into a crowd of Kent State University demonstrators in the United States, killing four and wounding nine Kent State students (Lewis and Hensley, 2009). If an authoritarian and non-democratic political regime were a necessary condition for the military crackdown, the violence at Kent State would have never happened.

Is the Kent State tragedy comparable to the Tiananmen tragedy? Although there are significant differences between the two, there are also many similarities (see Table 1.1).

Firing upon a crowd of student demonstrators was the ultimate outcome at both Tiananmen and Kent State. No matter how different the political systems in the United States and China, this basic fact cannot be denied. Since the same violent events occurred under both a democratic political regime and a non-democratic political regime, the existence of an authoritarian and non-democratic political regime cannot be a necessary condition for the violent events. An alternative explanation must exist—specifically one that explains why the same tragic events occurred in both a democratic country and a non-democratic country.

To summarize, all theories based on the typology of states (i.e., democratic vs. non-democratic regimes) neglect the important fact that the violence at Tiananmen was an unintended outcome. Moreover, they fail to provide any clarification about the conditions under which the Tiananmen tragedy occurred. What is wrong with the typology of states based on the concept of democracy, and why has it lagged behind the times?

Table 1.1 Comparison of the Tiananmen and Kent State tragedies

Comparison between	*Tiananmen tragedy*	*Kent State tragedy*
The goal of demonstration	Opposition to corruption and demand for democracy	Opposition to the Vietnam war and Nixon administration
The government's reaction to the demonstration	The government characterized the movement as a "planned conspiracy" and "turmoil"	Ohio Governor issued a statement calling protestors the worst type of people in America, and stating that every force of law would be used to deal with them
	The government implemented martial law in part of Beijing, and all demonstrations were banned	A state of martial law was declared, the National Guard assumed control of the campus, and all rallies were banned
The protestors' response to martial law	Asked government to "withdraw the troops and lift the martial law"	Protested the presence of National Guard on campus
	Formed barricades and built road blocks to prevent troops from entering the city Fraternized with soldiers and explained why they were blocking the army's entry	Set fire to the ROTC building adjacent to the Commons Often talked amicably with guardsmen
Major events before the shooting	Students blocked buses loaded with weapons; road blocks reappeared; hundreds of thousands of people blocked the troops	A rally was attended by 3,000 people, although university leaders printed and distributed 12,000 leaflets announcing the prohibition of rallies
Government's warning before the shooting	All government-controlled radio and television began broadcasting the announcement, throughout the city and the evening, repeatedly warning people not to go out into the street	A Kent State police officer was placed in a jeep with several guardsmen and driven across the Commons to tell protestors that the rally was banned and they must disperse
Protestors' response to the warning	Angry shouting; rocks and bottles thrown at Muxudi	Angry shouting; rocks thrown
Violent confrontation	The People's Liberation Army fired into a crowd of demonstrators	The Ohio National Guard fired into a crowd of demonstrators

This typology is based on the assumption that democracy is the best political system, and that it is of fundamental importance to focus on whether or not a regime is democratic. While this assumption is backed up by the rapid spread of democracy throughout much of the world, the typology is oversimplified and reality is much more complex. The typology of states based on the concept of democracy does not fit reality for the following reasons.

In general, separation exists between a state's political regime and its policies (economic, foreign, and social policy, etc.). In other words, for any country, whether its regime is democratic and whether its policies are intelligent, are two separate issues. For example, a country with the best political system may have a bad foreign policy, while a non-democratic government may have better economic and social policies. Without a doubt, a democratic regime has the highest degree of political freedom and political equality, but a regime high in democratic rankings does not guarantee that its officials will make intelligent decisions in relation to either foreign or domestic issues.

What determines a country's standing in the world? What makes a country decline or advance? It is obvious from the discussion above that the fact of its being democratic or non-democratic is not the only factor. For example, as Altman writes:

> The financial and economic crash of 2008, the worst in over 75 years, is a major geopolitical setback for the United States and Europe. Over the medium term, Washington and European governments will have neither the resources nor the economic credibility to play the role in global affairs that they otherwise would have played. These weaknesses will eventually be repaired, but in the interim, they will accelerate trends that are shifting the world's center of gravity away from the United States.
>
> (2009:27)

At the same time, China, with a non-democratic political regime, is achieving a stronger relative global position. Although democracy is the best political system, it seems that a democratic country without intelligent policies may decline, and a non-democratic country with intelligent policies may rise.

Since the democratic vs. non-democratic regimes dichotomy is behind the times, what would be the most useful way to classify countries? A new typology should be based on at least two concepts, or dimensions, of democracy and intelligence: whether a regime is democratic and whether its policies are intelligent. Both of these concepts are very important for understanding differences among regimes in the contemporary world. In the new typology, these two dimensions could have four different combinations: democracy and intelligence, non-democracy and intelligence, democracy and non-intelligence, non-democracy and non-intelligence.

While these four categories are not airtight, this typology's value lies in its ability to provide a new and useful way of thinking about the world. In studying the 1989 Chinese student movement, for example, scholars embracing this new way of thinking would never accept the assumption that the violence at Tiananmen was inevitable, simply because of China's undemocratic political system. Nor would they think that presenting the Tiananmen tragedy as an unintended outcome was politically incorrect. With the new typology, it would be possible to separate policy issues from political regimes and ask:

1 why the same tragic events occur in both a democratic country and a non-democratic one, and
2 whether a government's strategies in relation to a movement's development are intelligent or not, regardless of the nature of the political regime.

As mentioned earlier, there are two fundamental reasons for the fact that none of the existing studies has presented the Tiananmen tragedy as an unintended outcome. The first reason is that the typology of states based on the concept of democracy has fallen behind the times. When the world is changing, the typology of states must also change. The second reason is that the sociological perspective based on the sociological imagination has also become outmoded.

For many sociologists, the sociological imagination is a sociological way of thinking about the world. Emphasizing it lies at the core of sociological insight and intuition.

> C. Wright Mills defined the sociological imagination as the ability to see the societal patterns that influence the individual as well as groups of individuals. Sociology should be used, Mills argued, to reveal how the context of society shapes our lives; Mills thought that to understand the experience of a given person or group of people, one had to have knowledge of the social and historical context in which people lived.
>
> (Anderson and Taylor, 2007:4)

Mills' sociological imagination derived from Emile Durkheim, whose classic suicide study laid the foundation for modern sociological theory, in which social forces affect human behavior.

This theoretical orientation makes it is difficult to frame the idea that the outcome of the 1989 Chinese student movement was unintended. For scholars with a sociological perspective based on the sociological imagination, any outcome—either a military crackdown or a political solution—would be determined by "social forces," and it would never even occur to them to ask whether the violence was intended or unintended.

Tang Tsou, for example, provides us with knowledge of the historical context in which the Tiananmen tragedy occurred:

> Once we step back from our preoccupation with spectacular events, we can understand that Tiananmen signified nothing less than the latest in a series of failures in the twentieth century to rebuild the Chinese state, to restructure its society, and to establish a viable state–society relationship—this time in a wider, international environment that the Chinese leaders have not been able to understand. (1992:268)

Tsou examined the operative rules and norms of the Chinese Communist Party, and believed that, "All these rules and norms made it nearly impossible for the reformers to establish an early alliance with the students—the only strategy that could have prevented the Tiananmen tragedy, as we shall see" (1992:294). In Tang Tsou's study, the violence at Tiananmen was determined by social forces: the preceding development of Chinese history, including the operative rules and norms of the Chinese Communist Party.

Now assume that the emergence of social movements generates a new social system composed of two groups: a movement and a countermovement (in which the state is sometimes the countermovement). In explaining the outcomes of social movements, researchers with a sociological perspective based on the sociological imagination have focused on factors that are external to this new social system. For instance, Boswell and Dixon (1993) studied violent rebellion in 61 countries. In their regression equation, the dependent variable is deaths from political protests against the state, and the independent variables include class exploitation in manufacturing, income inequality, market crises, liberal democracy, etc. All of these independent variables are external to the newly created social system. While Tang Tsou (1992) provided the historical context in which the events at Tiananmen occurred, Boswell and Dixon (1993) described the social contexts in 61 countries in which violent rebellion took place.

Without a doubt, it is very important to have knowledge of the social and historical context in which events occurred and people lived. However, the utility of the sociological perspective based on the sociological imagination is facing greater and greater challenges. While social forces made the events at Tiananmen inevitable, what can we do to prevent such tragedies from happening in the future? When we know that in 61 countries correlations exist between deaths from political protest against the state and the prevailing social conditions, how can we minimize deaths within an unchanged social context? Studies based on the sociological imagination have not provided any answers to these questions. In other words, theory developed from these studies does not lead or lend itself to any application. Since Chinese history cannot be rewritten, for example, we must accept what happened at Tiananmen, and even that it will happen again in the future. Also, deaths from political protests against the state cannot be minimized in the 61 countries where they occurred until the social contexts of these countries change.

Hirshleifer and Glazer wrote that, "Theory is useless unless it leads to application" (1992:xiii). And, in sociology, there are many useless theories. "Social theory, by contrast, has been seldom used in relation to social policy. This is not a healthy state of affairs. One of the criteria for judging work in social theory is its potential usefulness for informing social policy" (Coleman, 1994:33). In sum, traditional sociological theory based on the sociological imagination has been elegant, but guilty of being irrelevant to the improvement of social conditions.

What is wrong with the sociological imagination? Why is theory that is based on it irrelevant to the improvement of society?

First, the sociological perspective based on the sociological imagination does not allow room for individual choices and public choices. It is true that the context of society shapes our lives, but the opposite side of the coin is that individuals choose how to lead their lives, and public choices generate the context of society. Sonia Sotomayor, the 111th Justice of the U.S. Supreme Court, was raised in a South Bronx public housing project. While American public housing has failed many minority kids, Sotomayor's success was based on her individual and family's choices. Unfortunately, the sociological imagination has emphasized only one side of the coin and has neglected the other side: the fact that individual and public choices generate the context of society. More importantly, study of the neglected side of the coin is the element that can produce theory leading to application.

When James S. Coleman studied American high schools, he found that students were influenced by social norms—social forces outside students' immediate, personal control.

> For example, norms regarding expenditure of effort toward school work develop among the students in most high schools. Often, these are norms opposed to working hard, norms which act to suppress effort. Sometimes, however, the norms go in the opposite direction, toward encouraging strong efforts and high achievement.
>
> (1994:33)

Then Coleman moved forward to challenge traditional sociological theory. He asked: "What are the conditions that determine the strength and direction of this norm" (1994:34)? It is this question that represents a new sociological way of thinking about the world. While Mills was interested in revealing how societal contexts, such as social norms, shape students' academic achievement, Coleman was interested in learning how to create social norms that encourage vigorous effort and hard work in American high schools. In order to develop a comprehensive theory of norm emergence that would inform the design of American high schools, Coleman's sociological perspective has focused on individual and collective choices.

Second, there is no room for assessing the social interaction among individuals or groups of individuals in the sociological perspective based on the sociological imagination.

> Social interaction [means that] . . . decisions are not made in a vacuum. Other people are likely to change their behavior in response to what you do. . . . When the likely responses of all the people affected are taken into account, the outcome is often much less attractive than it first appeared.
>
> (Hirshleifer and Glazer, 1992:2)

Since the interactions among individuals or groups of individuals determine social outcomes, such as the consequences of social movements or social norms existing in American high schools, it is very important to study how human beings interact with one another. Then why have the studies based on the sociological imagination failed to do so?

The problem lies in the research methods. The common method used in studies based on the sociological imagination is survey research, in which the unit of analysis is the individual or groups of individuals.

> Survey research was based on systematically gathered data from standardized interviews or questionnaires, often obtained via a randomly drawn sample from a well-defined population of individuals. . . . The data were combined via quantitative techniques, first cross-tabulations and later regression analysis and its extensions all directed to answering the causative question, accounting for individual actions or attitudes.
>
> (Coleman, 1994:30)

While survey research is very effective in showing how the context of society influences the individual as well as groups of individuals, it is an inadequate tool to study human interactions, for the following reasons.

As the unit of analysis in survey research, the individual or group of individuals is treated as independent. No connections or interactions exist among the units of analysis in the study. For example, in the general study of students' academic achievement at American high schools, the focus is on individual GPA (grade point average) as affected by the students' social background. Each student is treated as one independent unit of analysis, and there is no consideration of social interaction. In Coleman's 1994 study, by contrast, the unit of analysis is the high school—a social system—in which students, teachers, and parents interact with one another. Human interactions include school interactions, social movement interactions, family interactions, workplace interactions, etc. The first step in studying human interaction is the identification of the social system or the social setting as the unit of analysis. Unfortunately, most survey researchers have failed to do this; they have taken the individual instead of the social system as the unit of analysis.

There are some exceptions. In 1993, when Boswell and Dixon studied violent rebellion in 61 countries, they took the country as the unit of analysis, and it can be seen as a social system. But it was not possible for them to study rebellion interactions based on the data provided by their survey research. Human interaction starts with individual action, including individual preference and intention, and ends with system outcomes, such as the consequences of social movements or the social norms that exist at high schools, etc. Therefore, the second step in studying human interactions is the gathering of data or information at both the individual level and the system level, which makes it possible to study the characteristics of social interactions. The information that Boswell and Dixon gained from their survey research was about deaths from political protests against the state, class exploitation in manufacturing, income inequality, market crises, and liberal democracy. All of them characterize the social system of the country, but they are from only one level: the system level. In other words, their data did not provide any information about the individual level, such as information about the protestors' actions and the state's strategies, etc. Human interaction, however, cannot be analyzed without information from both levels.

Survey research has sometimes provided data from both the individual level and the system level. For instance, in the study of American high schools, researchers used both individual background differences and variations in system backgrounds—differences in school characteristics—to explain the variances in the student GPA. However, this has nothing do to with the study of human interactions, because the third step in studying human interactions is to concentrate on equilibrium in a particular social setting, such as the consequences of social movements or social norms existing at high schools. And here the explanation moves from the individual level to the system level, such as seeing how interactions among students, teachers, and parents generate social norms at high schools, or seeing how demonstrators' actions and state strategies generate the outcomes of social movements. But studies based on survey research do not concentrate on equilibrium because they have taken the equilibrium as given, and their explanations move from the system level to the individual level, such as seeing how society influences individuals or groups of individuals.

In short, the sociological perspective based on Mill's sociological imagination has become outdated. In regard to this perspective, the important fact that the Tiananmen tragedy was an unintended outcome has been totally neglected, and general sociological theory is once again irrelevant to the improvement of society. An alternative perspective has been presented by James S. Coleman in 1994, and the significant differences between Mill's sociological imagination and Coleman's sociological perspective are illustrated in Table 1.2.

Unintended Outcomes of Social Movements is based on Coleman's sociological perspective. The emergence of the 1989 student movement generated a new social system composed of two groups: a movement and the state. I take this new social system as the unit of analysis and focus on how the two groups interacted with each other and what characterized their interactions. More specifically, I present a condition that is both necessary and sufficient

Table 1.2 Comparison between Mill's sociological imagination and Coleman's sociological perspective

Comparison between	*Mill's sociological imagination*	*Coleman's sociological perspective*
Theoretical orientation	Durkheim social force	Max Weber and Talcott Parsons purposive action
The unit of analysis	Individuals or groups of individuals	Social system or social setting
The aim	To reveal how the context of society shapes people's lives	To explain how human beings interact with one another
Focus on	How properties of social systems have influenced individuals and groups of individuals	How properties of social systems have been produced by human interactions
	The explanation moves from system level to individual level	The explanation moves from individual level to system level
Research methods	Survey research	Formal modeling
Theory and application	Theory does not lead to application: knowledge of "what is" is irrelevant to improvement of society	Theory leads to application: knowledge of "what is" can be used to analyze the problem and suggest ways to achieve "what ought to be"

to create the Tiananmen tragedy: namely, a newly created social system with certain properties. I demonstrate that if this condition did not exist, the unintended outcomes of the movement would never have occurred, and that if it did exist, the tragedy would have occurred. In this way, I hope to inform people about the conditions under which unintended outcomes of social movements occur. Both social activists and governments can use this knowledge to find ways of achieving their intended outcomes.

The advantages of applying a game theoretical approach

Using Coleman's sociological perspective, I apply the game theoretical approach to the study of the 1989 student movement in China. Game theory is the study of interaction in competitive and cooperative environments, and game theorists seek to determine whether a given game can explain social phenomena. A game involves two or more players, where each player has at least two options. Each strives to obtain the greatest payoff possible, and the payoff achieved by one player depends not only on the option he chooses but also on the options chosen by the other players. There are several advantages to applying this approach to study of the Chinese student movement.

First, game theoretical models simplify reality in order to improve our understanding of it. After I have identified that the unit of analysis in my study is a newly created social system composed of two groups—social protestors and the Chinese government—a game can be used to represent the social system, and the two players in the game can be used to represent the two groups.

The model is stylized and it omits details. For instance, there were two aspects of the Chinese leadership in 1989:

1 the government was internally divided on how to respond to the demonstrations during the 1989 movement, and
2 the Central Committee of the Chinese Communist Party gave Deng Xiaoping the power to make all important decisions.

The game theoretical model omits the first aspect of Chinese leadership and includes only the second aspect—the one that is truly important to the movement's development. By omitting many irrelevant details, the model allows me to show, in a simple way, how the two groups interacted with each other to generate the eventual outcome of the movement.

Second, game theoretical analysis links the macro level of system functioning with the micro level of individual actions. According to Coleman, "The most formidable task of sociology is the development of a theory that will move from the micro level of action to the macro level of norms, social values, status distribution, and social conflict" (Coleman, 1994:33). I address this challenge by using a game theoretical analysis. In any given game, players (which can be individuals, groups, companies or countries, etc.) form the micro level, and the outcome of the game is generated at the macro level by players' interactions.

In the literature of social movements, there are many studies about interaction between movement and countermovement, or state. Here, however, I clearly identify for the first time that the emergence of social movements generates a new social system at the macro level, and that the outcomes and consequences of social movements function essentially as the equilibrium of that system. In order to understand this equilibrium, I use a dynamic game with incomplete information to capture basic aspects of the interactions between movement and state (or countermovement) that clearly demonstrate how micro-level actions intertwine to produce macro-level outcomes.

Third, game theoretical analyses are quite precise about the framework of institutional interaction, and demonstrate how institutions matter. According to Kreps, the institutional interaction framework is "about the connection between the choices of some individuals and the options available to and outcomes resulting for other individuals" (1990a:5). Blaug suggests that "the introduction of a game theoretical approach to economics has brought with it a new 'understanding' of what is meant by rationality and interdependence and equilibrium" (1992:240). The understanding of rationality is new because, in any given games, the outcomes resulting for a player depend not only on his own "rational choices" but also on the other player's decisions. Game theorists depict the institutional context of choices in various games. In *Unintended Outcomes of Social Movements*, for example, the framework of institutional interaction is composed of two parts:

1 the rules and constraints under which both the students and the Chinese authorities made their choices, and
2 the way in which the actions of both sides combined to produce the outcome.

More specifically, while other scholars focus more on an individual's intention in a given game, my thesis in *Unintended Outcomes of Social Movements* demonstrates the influence of institutions in a dynamic game. For example, I study why the preference structures of individual actors change during the dynamic game, how dynamics of the repeated game change players' payoffs, how having some limited information significantly influences players' strategy, and what determines individual actors' learning processes in a dynamic game. All of these elements are crucial to developing a complete and accurate explanation of the game's solution, but they can only be conceived of as external to the theory of intentional explanation.

Fourth, the modeling techniques in game theory offer a tool for analyzing dynamics. These dynamics become particularly important "when the actors make choices in period t that affect their set of feasible choices in a future time period t + t', where t' > 0" (Tirole, 1988:206). Two basic formal models are employed in non-cooperative game theory. The first one, a strategic form game, is used to study the static interactions between players. The second one, an extensive form game, is applied to the timing of actions that players may take. In *Unintended Outcomes of Social Movements*, I use the second formal model—an extensive form game—to draw a game tree as a way to think about the dynamic character of movement/state interactions.

Fifth, the language of game theory makes it possible for scholars of social movements to bring the issue of incomplete information into the study of outcomes. Information is very important to this study because rational actors make decisions based on the information they have about the costs and benefits of various courses of action. In game theory, the language of games with incomplete information, such as asymmetric or proprietary information, is well-suited to the study of a situation in which movement participants are unsure about the state's intentions. And it is equally well-suited to the analysis of how this uncertainty influences the participants' strategic choices. The concept of Bayesian equilibrium shows precisely how the concept of Nash's equilibrium may be extended to include this type of situation.

An important assumption

Game theory "provides general mathematical techniques for analyzing situations in which two or more individuals make decisions that will influence one another's welfare" (Myerson, 1991:xi). In this regard, the game theoretical approach I employ in this book is based on an important assumption: there is a single leader of the state who makes decisions on that side, and all of the demonstrators are wholly unified as a single individual who decides strategies on the other side. This single-player assumption is very important because game theorists are interested in explaining social phenomena as games with various payoffs. When discussion arises about these payoffs, the player on each side in the game has to be a unitary actor. Otherwise, no one understands whose payoff is being discussed. Although there is no doubt that almost all unitary actors are internally divided, the fragmentation of each side is not an issue in the study of the games.

Is the single-player assumption realistic? To answer this question, I introduce some specific circumstances during the Chinese student movement of 1989. First, was there a single leader of the Chinese government who made important decisions? Yes, there was. Although the Chinese government was internally divided during the 1989 movement, Deng Xiaoping was the only government official who made important decisions. Since 1978, there has been a high degree of consensus in the Central Committee of the Chinese Communist Party that all important government decisions should be made by Deng Xiaoping because he is one of the first-generation leaders of the Chinese Revolution, and he is experienced and highly respected.

Since 1978, no one in the government has ever successfully challenged Deng Xiaoping's position as decision maker. For example, the General Secretary of the Chinese Communist Party, Hu Yaobang, challenged Deng's authority in 1986, and was dismissed in January 1987. After his dismissal, the Central Committee of the Chinese Communist Party formally decided that all important decisions should be made by Deng Xiaoping.

The government was internally divided during the student movement of 1989, but a faction that disagreed with Deng had no power to make important decisions. Zhao Ziyang, the General Secretary of the Chinese Communist Party and a leader of this faction, was not even permitted to decide for himself to go to Tiananmen Square to talk with the dissidents, although 3,000 students went on hunger strikes and more than 500,000 people marched in the street to ask him to go. On 16 May, Zhao Ziyang openly explained why he could not go to Tiananmen Square: "All of us in the Central Committee have agreed that all the important decisions in the government should be made by Deng Xiaoping" (*China Daily,* 18 May 1989). In other words, only Deng had the right to decide whether the General Secretary, Zhao Ziyang, could speak directly with the students. On 19 May 1989, Zhao resigned because he knew it was impossible for him to challenge Deng's authority.

During the 1989 student movement, important decisions on the government's side included how to respond to the emergence of the student movement and whether to act on the government's initial threat after the demonstrators resisted it. All of these decisions were made by Deng Xiaoping. Although the Chinese military also experienced internal divisions—as did the civilian sector—no one in the army had the authority to defy Deng Xiaoping.

The assumption that there is a single leader of the state who makes important decisions is the consensus among political science and economics scholars. For example, in the literature of international relations, discussions on the role of military threats in deterrence and military crises are based on this assumption. And, in economics, study of the economic environment and strategic choices of governments is based on the same assumption.

In this book, I refer to the state as the subject of actions or initiator of behaviors based on the assumption that there is a single leader of the state who makes important decisions. I humanize the state, and try to avoid referring repeatedly to state authorities or agents or someone else that is a human being as the subject of actions. In the literature of international relations, it is common to refer to the state as the subject of actions or initiator of behaviors.

It is already assumed that there was one leader who made all the important decisions. Can all of the movement participants be wholly unified as a single individual, when there was internal disagreement among the students during the 1989 movement? According to *People's Daily* (which is one of the most important newspapers in China, and there is a high degree of consensus in China that *People's Daily* published the most objective report about the 1989 movement), more than 1,000,000 people spontaneously participated in the Chinese student movement of 1989 for democracy. They gathered at Tiananmen Square in May to support the hunger strike, and marched in the streets of Beijing. After the Chinese government declared martial law, about 1,000,000 people demonstrated in protest and prevented thousands of troops with tanks from approaching Tiananmen Square. After emergence of the movement, students represented less than 10 percent of the participants, and there was no organizational connection between student and non-student participants. Most non-student participants never even knew that there was internal disagreement among the students. In other words, the internal disagreement among students had no impact on the strategic choices of 90 percent of the demonstrators.

The assumption that all participants in the movement behave as a single individual is very popular in the study of rationality and large-number rebellious collective action (Moore, 1995), especially in the strategic analyses of this topic. In an assurance game (Taylor and Ward, 1982; Taylor, 1987; Chong, 1991), for example, one player is an individual participant, and another player represents all others. In other words, other participants behave as a single individual, although there is great internal disagreement among them.

A theory of interaction between social movements and state

I mentioned before that my goal in writing this book is to satisfy a public demand for knowledge about unintended outcomes of social movements—to present a condition that is both necessary and sufficient to create the Tiananmen tragedy, and to satisfy a professional demand to develop a theory of unintended outcomes of social movements. The emergence of the 1989 student movement generates a new social system composed of two groups: a movement and the state. The theory developed in this book focuses exclusively on the new social system and its particular properties that created a necessary and sufficient condition to produce the Tiananmen tragedy.

I coin four concepts in the book: anti-threat resistance, discriminatory two-level game, participants' tradeoff, and information gap. I then develop four propositions based on these concepts. Both the concepts and the propositions help us to understand the ways in which the newly created social system works. In general, the theory produces two important predictions:

1. the conditions under which threats made by the state are likely either to hinder the movements or to fail to do so, and
2. the conditions under which intended or unintended movement outcomes occur.

In Chapter 2, I present a brief history of the 1989 Chinese student movement. It is important for people to be aware that the movement had lasted for 50 days before the military crackdown. There were many unforgettable events and vivid features of the movement's development. It is incomplete, or even misleading, to provide any explanations of the Tiananmen tragedy without showing how the movement unfolded.

In Chapter 3, I examine three types of payoff structures between social movements and the state, and show how these different payoff structures can lead to different movement developments and outcomes. Of the three types, the game with incomplete information is the most important, because movement participants know nothing about the state's preferences and have their own beliefs about those preferences. In order to formalize the account of participants' response to the threat made by the state, I present a game theoretical model. In contrast to the simple account of the existing approaches, I demonstrate that anti-threat resistance is the Nash equilibrium in the game. This model specifies conditions under which threats made by the state during movement development are likely to hinder the movement or fail to do so.

In Chapter 4, I discuss the structure of a two-level game, which the state usually confronts after the emergence of the movement. Actions that are rational for the state in one game may be irrational for it in another game. More specifically, I distinguish the discriminatory two-level game from the intermediary two-level game. While the state must simultaneously make a move in two different games, the requirement for consistency between the two games is achieved by abandoning its optimal strategies, leaving the state with only sub-optimal strategies in the discriminatory two-level game. The signals disclosed by the state in choosing sub-optimal strategies significantly impact the movement's subsequent strategies. Playing these sub-optimal strategies also creates a situation that deprives the state of a means to disclose its real preferences.

In Chapter 5, I analyze the dynamics of repeated games, which capture the idea that players have to take into account the impact of their current strategies on their future choices and on other players' future actions. The most important of the dynamic characteristics is the fact that movement participants face a tradeoff between short-term gain and long-term loss. While the movement's payoffs change in the repeated games, I distinguish the state's "elastic" and "inelastic" responses to changes in the participants' payoffs. I prove that the more gains the participants reap in the short term, the more likely they are to lose in the long-term. In other words, if participants want to achieve their long-term goal, they must give up some short-term gains in the repeated games.

In Chapter 6, I focus on the final game, in which an information gap between the movement and the state causes bloody confrontation—an unintended outcome of social movement. There are two root causes of the information gap, but the fact that the state fails to choose strategies based on the credibility imperative is the most important one. I demonstrate how movement participants measure the state's credibility round by round in the repeated games, and how a decline in the state's credibility generates a domino effect. State authorities lose all means of showing their intention to use force and are unable to send a clear message to the movement. Lack of credibility paralyzes government officials when it comes to controlling the dynamics of the social movement. Bloody confrontation is the price that is paid for closing the information gap.

2 A brief history of the Chinese student movement for democracy (15 April 1989 to 4 June 1989)

15 April

- Hu Yaobang, a member of the Political Bureau of the Communist Party of China (CPC) Central Committee, and former General Secretary of the CPC, died suddenly of a heart attack at the age of 73.
- Many people were shocked to hear of his death. Hu was one of the major architects of China's reforms in the 1980s, and people considered him a great statesman and an outstanding leader.
- Most people associated Hu with their dreams of political reform, which they believed ought to go hand in hand with economic reform. Embezzlement, bribery, dirty tricks to make money, and a general disregard for the law indicated a high level of official corruption and moral deterioration in 1989 China. The opportunity to publicly mourn Hu provided a focus for expressing pent-up dissatisfaction.
- Big-character posters (a popular and traditional form of voicing protests and concerns in China) appeared at six or seven universities at Beijing: "Hu's death is a great loss for democracy and freedom." "Incompetent government, corrupted society, dictatorship, and depreciation of intellectuals: This is our society, the reality, and our tragedy" (*China Daily*, 16 April 1989).

16 April

- The people of Beijing spontaneously gathered at the Monument to the People's Heroes in Tiananmen Square to express their grief and dismay over Hu Yaobang's death.
- More big-character posters calling for greater freedom and democracy appeared on the campuses of 17 universities in Beijing.

17 April

- For the third consecutive day, thousands of students and residents in Beijing placed wreathes and Hu's portrait on the Monument to the People's Heroes. They also displayed posters to express their sorrow and anger: "The one, who shouldn't die, died, while those who should die still live on" (*China Daily*, 18 April 1989).
- The first demonstration, consisting of 600 student mourners, marched to Tiananmen Square at noon; the second march of 2,000 students took place that evening. On their way to Tiananmen Square, students chanted: "Down with corruption. Long lives democracy" (*China Daily*, 18 April 1989).

18 April

- After the second march arrived at Tiananmen, a student leader, Wang Dan, presented a seven-point petition to the Standing Committee of the National People's Congress, stating that:

  1 Hu Yaobang's contribution to China should be reassessed in a more positive light,
  2 the "anti-bourgeois liberalization" and "anti-spiritual pollution" campaigns against Western and capitalistic influences should be repealed,
  3 curbs on corruption and officials' privileges should be instituted,
  4 freedom of speech should be permitted,
  5 the right to protest and demonstrate should be protected,
  6 funding for education should be increased, and
  7 the rule against demonstrations stipulated by the Beijing City Municipality should be revoked.

  Several delegates from the National People's Congress came out to receive the students' petition.

- By nighttime, more than 10,000 students and other people took over Tiananmen Square to rally for democracy.
- Several thousand people left Tiananmen Square to go to Zhongnanhai, where the CPC headquarters and top government offices were located, and asked to speak directly to the nation's leaders.

19 April

- Demonstrators gathered in front of Zhongnanhai and tried to get in, but were stopped by guards. Some people threw bottles and shoes at the guards. Although the demonstration lasted for several hours, only a few minor cuts and bruises were incurred in scuffles. All police officers were unarmed and the incident was non-violent.

- A notice from the Beijing Municipal People's Government was read to the crowd in front of Zhongnanhai. The notice stated that the mourning of the late Hu Yaobang showed the people's love and remembrance of the dead, but added that certain persons wanted to incite trouble. The notice also urged the crowd to leave, and it gradually dispersed.
- Approximately 20,000–30,000 people gathered at Tiananmen Square, making this the third consecutive day of protests demanding significant changes in the political system: direct dialogue with government leaders, a free press, freedom of speech, and an end to corruption.
- Later in the day, several hundred students, followed by many onlookers, proceeded once more to party headquarters at Zhongnanhai. When they arrived, guards again prevented them from entering.

20 April

- The standoff in front of Zhongnanhai continued and the government again broadcast a warning asking people to disperse. While most people were persuaded to leave, about 200 students refused. A squad of unarmed police officers then came, dragged the students into a bus, and drove them back to campuses. The students were offended by the brutality of the police because some of them were kicked and beaten.
- A group of several thousand students marched to protest against the police brutality.

21 April

- Both the *People's Daily* (the CPC organ) and Xi Hua News Agency (China's official news agency) issued commentary. This was the first official response to the emergence of the Chinese student movement. Excerpts follow:

  > With shock and grievance, the party members and people across the country, including intellectuals and students, mourn the late Party leader Hu Yaobang and feel that they have lost a brilliant leader and a good teacher. Both their organized and voluntary memorial ceremonies expressed the sorrow in their minds.

> A gang of people were attempting to break through the Xihua Gate, the entrance to the headquarters of the Party Central Committee. That is absolutely forbidden. Anybody who seeks to infringe on the interests of the whole people is bound to be punished by law. Anyone who wants to plunder the party and the government institutions in the name of 'mourning' will be recorded as a 'sinner' in history.
>
> (*China Daily*, 22 April 1989)

- Forty-seven famous intellectuals sent an open letter to the Central Committee of the CPC, the Standing Committee of the National People's Congress, and the State Council (which is the equivalent of the State Department in the United States), expressing their support of the student movement.
- Some 40,000–50,000 students and residents of Beijing gathered in Tiananmen Square and continually mourned Hu Yaobang.
- The Beijing municipal government announced that Tiananmen Square would be closed off on 22 April during Hu Yaobang's memorial service. This made people very angry, because they believed that their right to mourn Hu Yaobang was being denied.
- By nighttime, the number of people gathered in Tiananmen Square exceeded 100,000, which made it impossible for security forces to clear the square.

22 April

- A memorial service for Hu Yaobang was held at the Great Hall of the People, located at west side of Tiananmen Square. Approximately 100,000 college students and other people stood outside in Tiananmen Square and listened to the service, which was broadcast over loudspeakers.
- After the memorial service, students gathered in front of the Great Hall of the People. They wanted to have a dialogue with Premier Li Peng. Three student representatives knelt at the stairs of the Great Hall holding a petition over their heads for more than 40 minutes. Students' emotions ran very high and they shouted: "Dialogue! Come out, Li Peng" (*China Daily*, 23 April 1989).
- Premier Li Peng did not come out, and the students were offended. Many of them cried and swore to continue their fight for democracy.

23 April

- The first unofficial student organization, the Autonomous Union of the Beijing Universities and Colleges (AUBUC), was established. One of the most important decisions made by the union at its first meeting was the launching of a citywide boycott of classes.
- Various activities took place on campuses: demonstrations, big-character posters, student meetings, petitions, etc. Through all of these activities, students called for political reform and targeted government officials' insensible response to students' enthusiasm for democracy.

24 April

- The classroom boycott began with an indefinite timeframe and included 60,000 students from 43 universities in Beijing.
- Grass-roots activities continued on Beijing campuses.

25 April

- Students from Qinghua University (one of the best universities in China) asked for a dialogue with high-level government officials, who accepted the request and scheduled a meeting.
- The students failed to elect representatives, however, and the meeting was canceled.
- The lower levels of government and other public bodies were informed of the gist of an editorial that would appear the next day in *People's Daily*, and were also told that a tough talk had been delivered by Deng Xiaoping, the most powerful man in China, on the basis of which the editorial was drafted.

26 April

- The editorial, entitled, "The Necessity of Taking a Clear-cut Stand against Disturbances," later known as the "April 26th editorial," was the lead editorial in the CPC organ, *People's Daily*. It characterized the student movement as a "planned conspiracy" and "turmoil," the essence of which was "basically to repudiate the leadership of the Chinese Communist Party and to repudiate the socialist institutions" (*People's Daily*, 26 April 1989). The editorial called upon the people to oppose the movement and predicted that the entire country would have no peace if the movement lasted. It was the first formal official response to the students' demonstration that sent a clear message that the government would no longer tolerate demonstrations.

- Both students and ordinary people were outraged by the editorial and the government's refusal to regard any of their requests as legitimate.
- AUBUC planned a large-scale demonstration for the following day to defy the editorial.

27 April

- In order to implement the April 26th editorial, Beijing municipal government officials promised the central government that they would do their best to stop student demonstrations. They put thousands of armed police officers on the street, forming seven blockade lines from Tiananmen Square to the northwestern suburb of Beijing, where most universities were located.
- To protest against the April 26th editorial, more than 30,000 students participated in a demonstration. The flow of marchers out of the northwestern suburb could not be staunched. While demonstrators confronted police officers face to face, they apparently did not want to clash with the police because, when they were stopped, they did not try to charge the police despite loud encouragement from some of the bystanders. They just kept shouting, "The people's police love the people" (*China Daily*, 28 April 1989). At the same time, the police were considerate and did not try to block the students. They just slowed down the marchers for some time and then let them pass.
- Some 200,000 Beijing residents stood on the sidewalks and watched the non-violent confrontation between students and police officers. They applauded the students.
- During the demonstration, students carried signs announcing their support for the CPC and their demands for greater democracy. They made it very clear that they opposed only the corruption, and were not conspiring to overthrow the socialist order, as they had been accused of doing in the April 26th editorial.
- The demonstration lasted 14 hours and was hailed by all students as one of the most successful, because of its enormous scale and the popular response it evoked.
- A State Council spokesman announced that the government would accept the students' demand for a direct dialogue with government officials.

28 April

- Student representatives prepared for a dialogue with government officials.

29 April

- A dialogue with central and Beijing municipal government officials took place. A total of 45 students from 16 Beijing universities and colleges attended the three-hour discussion. The main topics were the government's attitude towards student demonstrations, official profiteering, education, and the party's work style.
- The State Council spokesman said that Premier Li Peng had entrusted him to explain to the students that the April 26th editorial was directed against the unlawful acts of a few people, not at the masses of students. The spokesman also said that students' wishes to promote democracy and curb corruption were completely in line with government goals.
- Many students, who watched the televised dialogue on their campuses, challenged the legitimacy of the 45 student representatives who participated in the dialogue. These representatives had been selected by the All-China Students' Federation and the Beijing Students' Federation—China's official student associations—they had not been elected by their schoolmates.
- During the dialogue, a student requested that AUBUC, the unofficial student association established on 23 April, be given legal status. Although the union had been elected by the demonstrating students, the State Council spokesman said: "We do not recognize any illegal organizations" (*China Daily*, 30 April 1989).
- Students were outraged by the government's attitude toward AUBUC, which was the organization of their democratic movement.
- Some student leaders then held a news conference to announce that the dialogue was not what they required, and that AUBUC would not accept any dialogue with government officials arranged by China's official student associations.

30 April

- A group of 29 student representatives from 16 institutions of higher learning talked to Beijing municipal officials, including the mayor and the Communist Party Secretary of Beijing, who disclosed their personal incomes for the first time at the request of the students. Since this dialogue was arranged by Beijing's official student associations, it was not accepted by most students.

- Big-character posters appeared on campuses criticizing the two dialogues organized by government officials. At the top of a list of students' complaints and demands was official recognition of their autonomous organization instead of the party-controlled student unions already in place.

1 May

- Representatives of AUBUC held a press conference at Peking University, one of the top universities in China. More than 40 foreign correspondents participated in the conference, and thousands of students attended.
- The AUBUC spokesman announced nine documents. At the top was a document listing seven topics which the students wanted to discuss with government officials:

  1 reassessment of Hu Yaobang's contribution to China,
  2 recognition of the patriotism of the students and support for their demands for democracy,
  3 a criminal investigation into the police brutality of 20 April in front of Zhongnanhai,
  4 the prevailing problem of corruption,
  5 freedom of speech,
  6 an explanation of the state's negligence of education, and
  7 the failure of government policies in relation to economic reform.

2 May

- Students submitted their revised petition, including a list of demands, to the Standing Committee of the National People's Congress, the Party Central Committee, and the State Council.
- The petition presented the students' main preconditions for dialogue: the two parties should be completely equal in their positions vis-à-vis the dialogue and should sincerely work to solve problems; student representatives who participated in the dialogue should be acknowledged and selected by a majority of students; AUBUC must be permitted to contact students and organize the selection of

representatives; government participants should be familiar with state affairs and have decision-making power, etc.

- Students said that they would take to the streets again on 4 May if the government failed to respond to their demands by noon of 3 May.

3 May

- The state rejected the students' list of demands. A State Council spokesman expressed the government's willingness to continue dialogue with striking students but rejected the preconditions set forth by students in their petition. The government spokesman also considered AUBUC an illegal organization and described the students' main preconditions for talks as "unreasonable, emotionally impulsive, and menacing to the government in the form of an ultimatum" (*China Daily*, 4 May 1989).
- AUBUC planned a large-scale demonstration for the following day to protest the government's response to their petition.

4 May

- Tens of thousands of students from dozens of Beijing universities and colleges marched into Tiananmen Square demanding democracy, freedom of the press, and dialogue with government. The students, calling for a revival of the spirit of the "May 4th" Movement—China's first student movement, 70 years earlier—came to the vast expanse of Tiananmen Square soon after the official "May 4th" celebration ended.
- Student leaders declared an end to the two-week-long class boycott, and called on all the striking students in Beijing to resume classes. However, they also reiterated their demand for a dialogue with the government and vowed that they would retain the right to take further action.
- More than 200 journalists from the Beijing-based mass media joined the students in an unprecedented protest calling for freedom of the press.
- The demonstration attracted many bystanders, but the total number appeared to be smaller than on 27 April.
- "The demonstrators in no way mean to oppose the fundamental system of China, but to correct the errors in the work of the party

and the government." Zhao Ziyang, General Secretary of the CPC, said, while meeting with governors of the Asian Development Bank during its 22nd annual meeting in Beijing (*China Daily*, 5 May 1989). Zhao Ziyang also said that reasonable demands from the students should be met through reforms and various other reasonable democratic and legal means, and declared that the most important immediate needs were calm, reason, restraint, and order.

5 May

- Most of Beijing's striking students responded favorably to party Chief Zhao Ziyang's speech. They liked its mild tone, which they believed showed sincerity and reason on the government's side.
- Students started going back to their classes in accordance with the declaration by student leaders during the May 4th demonstration, although some demanded that they should continue their two-week-old boycott. Students split over whether or not to return to their classrooms.
- Zhao's talk on 4 May seemed to have had a calming effect on most students, and the democratic movement seemed to be ebbing.

6 May

- Many students at Peking University continued to stay away from classes despite the student leaders' announcement on 4 May that classes would resume. Students who refused to go back to classes insisted on a continuous protest, because they did not think that their goals had been reached. They insisted that government officials have a dialogue with student-movement delegates instead of representatives from the official student unions.
- Four student representatives handed in a petition demanding talks. The petition carried signatures from 24 Beijing universities and colleges and was addressed to the CPC Central Committee, the Standing Committee of the National People's Congress, and the State Council. It called for "sincere, constructive and open dialogues on issues including the current students' patriotic movement for democracy, furthering political and economic restructuring, and promoting democracy and legislation" (*China Daily*, 7 May 1989).

7 May

- A dialogue was held between the ministers of the Agricultural Ministry and students from Beijing Agricultural University and Beijing Agro-Engineering University.
- Student leaders from Peking University decided to continue to boycott their classes for another week, despite the fact that students at other universities and colleges had returned to their classrooms.

8 May

- The four student representatives who had handed in the petition received a pledge of more dialogues from the director of the department who was charged with handling people's representations vis-à-vis the general office of the CPC Central Committee and the State Council.

9 May

- Staff members of the *China Youth News* handed a petition to the All-China Journalists Association. The petition was signed by over 1,000 Beijing journalists and called for a dialogue with party and government officials to establish freedom of the press. The petition also criticized censorship in reporting the student movement.
- More than 1,500 students demonstrated in support of the journalists' petition.

10 May

- CPC leader Zhao Ziyang stressed the importance of reforming China's political structure, but said that reform needed to proceed gradually. Building a legal system and developing democracy constituted the central tasks for political reform, Zhao Ziyang told a delegation of the Bulgarian Communist Party.
- Thousands of college students took to their bicycles in a demonstration in Beijing, calling for freedom of the press and immediate dialogue with the party and government officials. They chanted slogans against what they deemed to be subjective reporting of recent student activities. This action, organized by AUBUC, was the largest student demonstration since the massive one on 4 May.

11 May

- Hu Qili, a member of the Standing Committee of the Political Bureau of the CPC Central Committee, met with editors and reporters of *China Youth News* who had signed and handed in the petition on 9 May. Hu listened to the opinions voiced by participants.
- The department charged with handling people's representations to the general office of the CPC Central Committee and the State Council answered students' questions about the specific arrangement of dialogues with government officials.
- While it seemed that the movement had reached a stage where demonstrations and classroom boycotts would be replaced by various dialogues with government officials, student leaders saw a unique and unplanned opportunity approaching: Soviet leader Mikhail Gorbachev was coming to China to hold summit talks with senior Chinese leader Deng Xiaoping from 15 to 18 May. The students respected Gorbachev for the political reforms he had initiated in the Soviet Union. They also knew that the CPC attached great importance to Gorbachev's visit, and that the threat of disruption to this event could be used as leverage to pressure the party.
- Student leaders planned to begin a hunger strike on 13 May at Tiananmen Square, where the welcoming ceremony for Gorbachev had been scheduled. They believed that this plan would not only become a turning point by bolstering students' morale, but also raise the bargaining power of the students to new heights.

12 May

- Wan Li, Chairmen of the Standing Committee of the National People's Congress of China (NPCC), announced that the Standing Committee of the NPCC would hold its eighth meeting, lasting for a week, starting on 20 June. Major issues on the agenda included clearing up corruption, surveying the demands that students proposed in their demonstrations, and promoting democracy through the legal system. Most students and ordinary people responded favorably to Wan's announcement.
- The four student representatives who had handed in the petition demanding talks on 6 May brought a list of students who would participate in the prospective dialogues with government officials, and presented this list to the department charged with handling people's

representations to the general office of the CPC Central Committee and the State Council. Both students and government officials were then involved in the process of preparing for the dialogues.

- At a news conference, a State Council spokesman urged students to refrain from staging a mass demonstration during the Sino–Soviet summit.
- Students prepared for a hunger strike.

13 May

- The director of the department charged with handling people's representations to the general office of the CPC Central Committee and the State Council informed the student representatives, who had handed in the petition on 6 May, that the leaders of the CPC Central Committee and the State Council would continue a dialogue with students and other people on 15 May.
- Zhao Ziyang, General Secretary of the CPC, met with worker representatives to discuss political reform and the Chinese student movement.
- Premier Li Peng had a dialogue with steel workers in Beijing to talk about China's economic situation and the student movement.
- Approximately 1,000 students gathered in Tiananmen Square and began fasting to protest what they described as "false charges against their previous activities" and "the government's delay to conduct substantive dialogues" with student representatives (*China Daily*, 14 May 1989). The fasting students, who wore white-and-red headbands, gave either the names of their universities or flaunted political slogans, and said that theirs was a patriotic and democratic movement—not "disorder," as claimed by the authorities.
- Hundreds of thousands of people gathered at Tiananmen Square. The hunger strike gave the student movement a focal point that brought about an unprecedented outpouring of sympathy from the citizens of Beijing, young and old, rich and poor, highly educated and semi-literate. It raised the student movement to new heights.
- Yan Mingfu, a member of the Central Committee's Secretariat and head of the United Front Department, met with student leaders to arrange a dialogue for the following day.
- Yan Mingfu reported on his meeting with student leaders to the top officials of the CPC Central Committee, who accepted students' demand for a dialogue.

- In the evening, the Beijing Emergency Medical Centre sent two ambulances to Tiananmen Square to be ready for rescue work at any time.

14 May

- Li Tieying, a member of the CPC's Political Bureau and minister in charge of the State Education Commission, Chen Xitong, Mayor of Beijing, and Li Ximing, Communist Party Secretary of Beijing, went to Tiananmen Square early in the morning to persuade fasting students to end their demonstration and return to their schools.
- More students joined the hunger strike.
- More than 10,000 students marched in Tiananmen Square to support the fasting students.
- There were about 10,000 onlookers at Tiananmen Square.
- A dialogue between student representatives and high-ranking party and state officials, Li Tieying, Yan Mingfu, and Wei Jianxing took place as about 1,000 students continued their hunger strike. This was the first time that the government had engaged in a dialogue with student representatives who had been elected by unofficial student organizations. The meeting between government officials and some 50 student representatives from about 30 Beijing universities and colleges began in a sincere and reasoned atmosphere. The students' main demands and concerns were that the dialogue be broadcast live to the fasting students in Tiananmen Square, and that the government renounce its allegation that students' actions during the mourning for late party leader Hu Yaobang were creating "chaos."
- The dialogue went into recess after three hours, when it became apparent that the two main demands of the students would not be met.
- Twelve leading intellectuals issued the most noteworthy statement made during the student movement, affirming their belief that the student associations were "legal organizations" and should be recognized by the government. They also demanded that the party openly declare the student upheaval as a patriotic, democratic movement, and that the government take no action against students after the dispute was settled. They expressed their opposition to any use of violence against the students. At the same time, they asked fasting students to end their demonstration so that the Sino–Soviet summit could take place smoothly.

- More than ten fasting students collapsed. The Beijing municipal government immediately sent doctors and ambulances to the square, and the students were hospitalized.

15 May

- China's official student associations, the All-China Students' Federation and Beijing Students' Federation, arranged another dialogue between high-ranking officials Li Tieying and Yan Mingfu, and 50 student representatives. The students' main demand was still a retraction of the allegation that their actions during the mourning for late party leader Hu Yaobang were creating "chaos." The dialogue went into recess after three hours when the students' main demand could not be met.
- As the hunger strike entered its third day, the number of fasters kept expanding. On the first day, there had been about 1,000; on the third day, there were about 3,000. The number of students who fainted from hunger also increased. By 6:30 p.m., as many as 103 strikers had collapsed after more than 50 hours of fasting.
- About 1,000 volunteer medical personnel and students from the Union Hospital, one of the best hospitals in Beijing, were on hand to give first-aid to hunger strikers.
- In an open letter to party and government leaders, faculty members from Peking University urged them to meet student representatives as soon as possible for substantive dialogue. The letter, which was signed by more than 500 professors, including many respected scholars, also urged the government not to delay in accurately evaluating the month-long student movement.
- About 40,000 intellectuals walked in groups to Tiananmen Square to support the fasting students. Journalists, writers, and social and natural scientists were among the supporters.
- More than 100,000 people milled around Tiananmen Square to support the fasting students.
- Gorbachev arrived in Beijing. The welcome ceremony had to be moved to the airport, which caused the Chinese government to lose face.
- Although Beijing's Red Cross officially assumed all medical care for the students at Tiananmen Square, thousands of volunteer medical workers from 32 hospitals also offered first-aid to the exhausted hunger strikers.

16 May

- The sound of ambulance sirens filled Tiananmen Square as the ranks of students participating in the hunger strike swelled to 3,147. After more than 70 hours without food, 617 student hunger strikers had fainted and been taken to emergency rooms and 247 were hospitalized.
- Yan Mingfu, a member of the Secretariat of the CPC's Central Committee, who was also the minister in charge of the Party's United Front Department, spoke to the hunger strikers and other students in Tiananmen Square, and told them that the authorities would not retaliate in any way against anyone who joined or supported the current student movement. He said he was making the pledge on behalf of the CPC Central Committee, and added that a meeting of the party at the highest level would be convened as soon as possible to study students' demands. Pending that meeting, he said, students had the right to carry out further strikes and demonstrations, but he urged them to stop their hunger strike.
- Student leaders endorsed Yan's plea. They said they believed he was sincere and asked the other students to believe them. But most representatives of the striking students voted to continue their hunger strike until a clear-cut agreement with the government was reached.
- In support of the students, between 500,000 and 600,000 people went to Tiananmen Square. Among them were teachers, professors, writers, journalists, and workers. Crowds were still streaming towards Tiananmen Square around 7:30 p.m., blocking traffic along major streets.
- The directors of 40 Beijing hospitals gathered at the city's emergency care center to discuss how to take care of the student hunger strikers at Tiananmen Square. More than 40 doctors and nurses, with ambulances, had stayed with the fasting students in the square around the clock since the early morning of 15 May.
- Senior Chinese leader Deng Xiaoping and visiting Soviet leader Mikhail Gorbachev held their long-awaited summit meeting in Beijing, marking the normalization of state and party relations between the two countries. The two leaders did not say a word about the ongoing student demonstrations, which had forced several last-minute changes in Gorbachev's scheduled program.
- Zhao Ziyang, CPC General Secretary, told his Soviet counterpart, Mikhail Gorbachev, that senior leader, Deng Xiaoping, was still

China's helmsman when it came to important issues. Although the 85-year-old Deng had voluntarily relinquished all of his positions in the party in 1978, Zhao said, "All the comrades hold that in the interests of the party, we still need his wisdom and experience" (*China Daily*, 17 May 1989). Since the 13th National Party Congress in 1987, he said, the party had always reported to Deng and consulted him on major issues.

- China was prepared to improve its record on freedom, democracy, and human rights in the course of its political reform, Chinese Premier Li Peng said in talks with visiting Soviet leader Gorbachev.
- Presidents of ten Beijing universities signed an open letter in which they expressed anxiety about the situation in Beijing. The presidents said:

> We are deeply concerned about the health and lives of the students, and the future of our country. On no account do we want to see the course of reform—the policy of opening up to outside of the world, and modernization—being disrupted or discontinued.
> (*China Daily*, 18 May 1989)

They hoped that the major leaders of the party and the government would come out as early as possible to see the students and conduct a dialogue with them.

- Twelve student hunger strikers refused to drink a single drop of water. The volunteer medical team from the Beijing Emergency Medical Centre used water-soaked cotton to rub their lips as much as possible, and doctors checked the pulse and blood pressure of the 12 students every half hour.

17 May

- On behalf of all the members of the Standing Committee of the Party Central Committee's Political Bureau, General Secretary Zhao Ziyang appealed to the hunger-striking students in a written message delivered early in the morning. Zhao said that the CPC Central Committee and the State Council affirmed the students' patriotic spirit in calling for democracy and law, opposing corruption, and striving to further the reform. "We also hope that all students will exercise calm, reason, restraint, and order; take the interests of the whole country into account; and safeguard stability and unity," he said (*China Daily*, 18 May 1989).

- One student leader told *China Daily*: "Zhao's statement did not touch on our two basic demands: acknowledge our movement to be patriotic and democratic, and hold substantive discussions with us. So we are not satisfied and we will keep on fasting" (*China Daily*, 18 May 1989).
- More than 1,000,000 Beijing citizens from all walks of life spontaneously demonstrated in Beijing in support of some 3,000 fasting students in Tiananmen Square, while talks between the students and government remain deadlocked. The demonstrators included journalists, factory workers, high school students, government office workers, bank clerks, and non-combatant officers of the People's Liberation Army. The march was the largest mass demonstration since the death of Hu Yaobang a month earlier.
- Four of China's nine smaller democratic parties that were allied with the CPC urged Communist Party Secretary, Zhao Ziyang, and other top leaders, to go to Tiananmen Square immediately and open a dialogue with demonstrating students so as to bring an end to their hunger strike as soon as possible. Their appeal was echoed by the China Federation of Literary and Art Circles, the All-China Women's Federation.
- More than 1,000 volunteer medical workers in white gowns brought badly needed saline solution and dextrose to the fasting students and administered first aid. By 6:00 p.m., 1,768 students had fainted and been taken away to hospital. Li Tieying, a member of the CPC's Political Bureau and the minister in charge of the State Education Commission, visited the students in hospital.
- Marches in support of Beijing's hunger-striking students were held in more than 20 of China's provincial capitals. In Shanghai, some 60,000 students staged a sit-in in front of municipal headquarters and in the People's Square. Of these, 500 had started fasting.
- Foreign business people expressed both surprise and concern as more than 1,000,000 people poured onto the streets of Beijing in support of the student hunger strikers. The deputy general manager of NTT (Nippon Telegraph and Telephone Corporation) Beijing's Representative Office told *China Daily* that he and his colleagues were initially very nervous about the demonstrations. But when they found many marchers in a jubilant spirit, their fears began to ease. He said that he went to Tiananmen Square and was very moved by what he saw. People voluntarily maintained traffic order so that ambulances could move freely in and out of the hunger-strike zone.

"We are very concerned how the movement will end," he said, adding that he hoped nothing worse would happen (*China Daily*, 18 May 1989). (NTT's business in China included providing communication technology and equipment to the Ministry of Posts and Telecommunications, the Railway Ministry, and the State Information Center.)

18 May

- CPC General Secretary Zhao Ziyang, Premier Li Peng, and other members of the Standing Committee of the Political Bureau and the Secretariat went to hospitals at 5:00 a.m. to comfort hunger strikers.
- Thousands of Beijing students were in their sixth day of fasting, and more than 2,000 had fainted and been taken to hospital. Beijing citizens sobbed as they chanted slogans from the tops of big trucks in support of the students, watched ambulances heading away from Tiananmen Square carrying exhausted students, and saw many students lying motionless on the square in threes and fours. When medical workers put unconscious students into ambulances at the rate of one every five minutes, Beijing residents could no longer keep back their tears. In addition, many university professors came to Tiananmen Square to see their beloved students and were seen wiping away tears with their handkerchiefs.
- Demonstrating students gathered around more than 90 public buses in which hunger strikers had taken shelter after overnight showers drenched Tiananmen Square. Mobilized by the Beijing Red Cross early that morning, the buses protected students from the thunderstorm that occurred during the day. Factories donated umbrellas and plastic cloth, and residents of Beijing donated food and money to demonstrating students. A college student who participated in the hunger strike said: "I was moved to tears when it was raining. The doctor and nurses covered us with plastic cloth while they were standing in the rain" (*China Daily*, 19 May 1989).
- More than 1,000,000 people demonstrated for the second day in support of the hunger strikers. For the first time, workers came out in large numbers, not just to share in the students' heroism, but to speak on their behalf. The Beijing Workers' Autonomous Union was formally established.
- The *People's Daily*, China's leading newspaper, continued to publish appeals to party and government leaders calling for immediate

constructive dialogue with students. The people who sent appeals included 18 professors from China Science and Technology University, and 20 well-known writers, managers, and directors from Beijing's ten factories.

- A dialogue between Premier Li Peng and leaders of the striking students took place in the Great Hall of the People in Tiananmen Square. Li emphasized that the first priority was to rescue and guarantee the lives of the hunger strikers. But he did not accept the students' key demand for an overall rehabilitation of their movement, saying that even if their original wish had not been to create chaos, turmoil had, in fact, occurred. "Beijing is on the brink of anarchy," he said (*China Daily*, 19 May 1989). During the meeting, some students questioned the Premier's sincerity and threatened to walk out. The dialogue failed to resolve the rift between the students and the government.
- Soon after their meeting with Li Peng, student leaders announced a list of new demands in which they asked that emergency meetings of the party's political bureau and the National People's Congress be convened in order to officially affirm that the student movement was a great, widespread, patriotic, and democratic movement. They also said that legal sanctions should be instituted against government officials responsible for the continued ignoring of the people's just demands—a disregard that had led to the spreading of disaffection.
- Some political columnists of Chinese newspapers held an emergency meeting to study the country's critical political situation and declared that the central leadership should not, at any cost, lose its last opportunity to resolve the student petition. They believed that the leadership and all segments of society should learn how to seek compromise, build consensus, and work toward the country's political modernization in democratic and legal ways.
- By 6:00 p.m., Beijing's Emergency Medical Centre reported that 3,504 hunger students had been taken to emergency rooms and that 2,457 were hospitalized during the six days.
- Tensions rose at Tiananmen Square during the night because the dialogue with Premier Li Peng turned into a confrontation that left neither side room for future maneuvering.

19 May

- CPC General Secretary Zhao Ziyang and Premier Li Peng visited hunger-striking students in Tiananmen Square shortly before dawn.

When Zhao and Li approached the students' buses, many students applauded and others stretched their arms out of windows and shook hands with the leaders. With tears in his eyes, Zhao appealed to the hunger strikers to end their seven-day fast: "You all want our country to become better," he said, adding that the problems they raised would eventually be solved (*China Daily*, 20 May 1989). He also said that the situation was complicated and that a process was needed to resolve the problems. Zhao signed students' hats and shirts.

- Huge numbers of students from other parts of the country had flocked into Beijing since 16 May to support the hunger strikers in Tiananmen Square. Their numbers reached a peak on 19 May, when 50,000 arrived. Many had traveled free of charge with the help of railway workers.
- In the late evening, at a huge assembly attended by party, government, and army officials, Premier Li Peng declared that tough action, with the backing of the army, was needed to quell the "turmoil." He said that it had become more and more clear that the few people who attempted to create turmoil simply wanted to reach their political goals: negating the leadership of the CPC and the socialist system. Their purpose was to overthrow the people's government, elected by the National People's Congress, and destroy the people's democratic dictatorship. He further stated that they stirred up trouble everywhere, established secret ties, set up illegal organizations, and forced the party and the government to recognize them. He called on the entire party, army, and nation to act immediately to stabilize the situation.
- Communist Party General Secretary Zhao Ziyang refused to take part in the meeting because he had refused to implement martial law. On 24 June, Zhao was formally relieved of all his posts in the party by the Fourth Plenum of the 13th Central Committee. The official charge against him was that "in the critical hours of the life and death of the party and the state, comrade Zhao Ziyang committed the error of supporting the turmoil and splitting the party" (*China Daily*, 20 May 1989).
- At 9:00 p.m., student leaders announced at Tiananmen Square that the hunger strike had ended.

20 May

- Most student fasters had ended their hunger strike in Tiananmen Square by 3:00 a.m. Some had gone back to their schools and homes as the loudspeakers repeatedly broadcast speeches made by Premier Li

Peng. However, the square was still packed with thousands of students, including some who continued their eight-day hunger strike.

- In the early morning, soon after the central government announced over television that troops had been ordered into the capital to quell the "turmoil," students and local residents began setting up roadblocks at major intersections around Beijing in an attempt to stop truckloads of soldiers. They turned over buses, trolley cars, garbage cans, road dividers, and whatever else they could find in the vicinity to serve as roadblocks. At least 273 buses and trolley cars were used as barriers, according to Zhang Baifa, Vice Mayor of Beijing.
- The army convoys were on their way to downtown Beijing in the early morning hours to enforce martial law when they were blocked. Students and ordinary citizens called the troops the people's own army and offered soldiers food for breakfast.
- The State Council of the People's Republic of China issued an order early in the morning, signed by Premier Li Peng, to execute martial law throughout most of Beijing. The order stated that serious turmoil had taken place in the city, and that social stability, people's normal lives, and the social order had been disrupted; in order to put a decisive stop to the unrest, safeguard social tranquility in Beijing, safeguard the life and property of the citizens, protect public property, and ensure the normal function of the central departments and Beijing's municipal government, the State Council had decided to implement martial law in part of Beijing at 10:00 a.m. on 20 May 1989.
- Meanwhile, Beijing's mayor, Chen Xitong, signed three orders established by the municipal government. Excerpts follow:

> Under the martial law, demonstrations, petition, class boycott, work stoppage, and other activities of amassing people to impede the normal order, are banned.
>
> People are banned from using any means to create and spread rumors, establish ties, make public speeches, or distribute leaflets to incite social turmoil.
>
> Under martial law, public security officers, the armed police, and army on duty have the right to use every possible means to stop any of the above-prohibited activities, should they occur.
>
> (*China Daily*, 22 May 1989)

- Enforcement of martial law obviously failed, as more than 200,000 students began a peaceful sit-in protest at Tiananmen Square after the hunger strike had ended.

- Approximately 1,000,000 people poured into the streets of Beijing to protest against martial law. The overwhelming majority of demonstrators' slogans were directed against the principal leader of the State Council. Other slogans included, "Withdraw the troops and lift the martial law," and "Safeguard the constitution and guarantee human rights" (*China Daily*, 21 May 1989).
- In the evening, local residents spontaneously gathered at all the major entrances into the city to form barricades and build roadblocks in a bid to prevent military troops from entering.

21 May

- The headquarters for the enforcement of martial law in Beijing asked for understanding and cooperation from citizens, especially "the patriotic students." A notice issued by the army command said that the troops were by no means in the city for action against the students or citizens, but would help local police and the military police execute the order for martial law. At the time the notice was going to press, troops were already waiting at the city's outskirts for orders to move. The notice also said that the army command would strictly abide by various disciplines and ensure that troops were thoroughly briefed about "loving the capital, loving the capital's people, and loving young students" (*China Daily*, 22 May 1989).
- The 2,000 student hunger strikers began to eat, and joined tens of thousands of others in a peaceful sit-in in Tiananmen Square asking for the lifting of martial law.
- More than 1,000,000 people took to the streets protesting persistently against martial law. They came to Tiananmen Square and crowded onto the adjacent Changan Boulevard, nearby streets, and major road junctures.
- When military trucks were blocked by supporters of the student movement, the people of Beijing enthusiastically tried to explain to the soldiers why they were blocking their entry into the city. "We're not your enemies—let's be comrades together," they said (*China Daily*, 22 May 1989). They gave soldiers newspapers, cigarettes, and soft drinks. One old lady, who had actually lain down in front of a truck and said, "If you want to move, drive over me," was later seen bringing the soldiers—many of whom were the same age as the student strikers—steamed bread and drinking water, and chatting with them like a mother with her children (*China Daily*, 22 May 1989).

- Eight top generals from the People's Liberation Army urged the government not to use force against students.
- Zhang Baifa, Vice Mayor of Beijing, appealed to the public to unblock urban roads because necessary supplies to the capital were being adversely affected. In a television broadcast, he said that Beijing's coal, liquefied gas, and electricity were in short supply. Vegetables, milk, eggs, and other food staples could not be transported into the city's center, and garbage could not be carried out.
- Rumors spread in Beijing that the troops were going to suppress the students during the night. Hundreds of thousands of residents flooded into the streets and Tiananmen Square that evening. They formed barricades and built roadblocks. Students in the square spent an anxious night.
- In the evening, several student representatives went to the residences of Marshals Nie Rongzhen and Xu Xiangqian to submit a letter asking them to stop the military plan to suppress the students. Both marshals clarified that quelling the students was a rumor, and that the troops had come to Beijing to impose martial law to safeguard the social order and stability of the capital. Students, they said, should immediately withdraw from Tiananmen Square in order to restore order, national dignity, and urban life—as well as their own health and studies.
- Student leaders debated whether to continue the sit-in strike or to call it off. The latter would start a long-term struggle for democracy. They could not reach any agreement.

22 May

- In the early morning hours, barricades were removed from the 24-kilometer-long Second Ring Road, the main highway encircling central Beijing.
- Student groups at every barricade helped remove the barriers to allow regular traffic movement in the morning. They worked very hard to do several things at the same time: keep the peaceful sit-in in Tiananmen Square, fraternize with the soldiers, and help to maintain normal order in the city's life.
- The Beijing Municipal Government and the People's Liberation Army martial law enforcement headquarters issued a public announcement calling for the restoration of normal order in the capital as soon as possible.

- Tiananmen Square, occupied by thousands of students, was comparatively quiet three days after the imposition of martial law. Although many Beijing college students went back to their campuses for a rest, tens of thousands of students from other parts of the country remained in the square. Many Beijing citizens and factory workers brought in fast food and meals. A number of ambulances were still on hand, and volunteer contingents of doctors and nurses still served the sick and occasionally distributed medicines to the students.
- The Beijing railway station began in earnest to try to get the students to return home. Railway officials set up several counters in Tiananmen Square to offer students a free ticket home upon presentation of a valid student identity card. About 13,000 had left Beijing after joining the demonstrations for a day or two, but about 70,000 college students from other parts of the country remained in the capital. The free-ticket offer was to last from May 22 to May 25.
- Several thousand Beijing intellectuals, including many journalists, marched to the square that afternoon to show their support for the students.
- Wan Runnan, president of the Beijing Stone Group Corporation and an intellectual-activist, proposed at a meeting with almost 100 student leaders that they "should not lose the opportune movement to withdraw from the square, but should take the initiative to go back to the campuses and to persist in the struggle there" (*China Daily*, 23 May 1989). However, Wan stated that there were certain questions and demands that would accompany the withdrawal. First, the troops should leave. Second, martial law should be canceled. Third, Premier Li Peng should step down and Deng Xiaoping, Yang Shangkun, and the other old men should retire. Finally, and most importantly, Zhao Ziyang should be given support. However, the students did not accept Wan Runnan's proposal because they did not want withdraw from the square.
- Without consulting other student leaders, Wuerkaixi, a chairman of AUBUC, asked that students evacuate the square. He was dismissed from his post and a collective leadership was established.
- Field army units from outlying provinces were still stationed outside the central city, blocked by hundreds of thousands of Beijing citizens.
- China's central television station interviewed some officers of a People's Liberation Army martial law enforcement unit who had arrived in Beijing for three days. Excerpts follow:

> Our troops are unable to move into Tiananmen Square. As we are stopped by the people, we are unable to move forward. Our

> superiors are letting us remain here while awaiting further orders.
>
> Since we arrived, we have seen that the college students are quite reasonable. Like us, they try to avoid confrontations. Since we arrived, we have also acted according to the demands of the Central Military Commission to try to avoid clashes. But the people do not understand us. Sometimes they uttered unfriendly words. We have instructed our troops to ignore all this and to try their best to avoid clashes. In this regard, the college students have also helped us do a lot of work. So far, our troops have not run into any conflict with the students and the people.
>
> Our soldiers have stayed in the army trucks. Despite the heat and hunger, they all strictly observed discipline. Today, our superiors let us retreat. Our soldiers do not carry bedding or rain gear, and now they all lie on the ground. Just now, I made an inspection tour and it was painful to see. I told our officers to wake up the soldiers after 4:00 in the morning because it was getting cold and damp on the ground, and to let them sit up before returning to the trucks.
>
> Residents of Beijing took care of our food and water. They brought us bread, steamed bread, porridge, and salted vegetables.
>
> (*China Daily*, 24 May 1989)

23 May

- Thousands of students continued their sit-in at Tiananmen Square. They were caught in a thunderstorm and had to seek shelter in the afternoon. Beijing citizens later responded enthusiastically to an urgent Red Cross call for warm clothes and covers to help the thousands of students get through the night.
- Some 1,000,000 people took to the Beijing streets in the biggest demonstration since the imposition of martial law. In addition to college teachers and students, people from journalistic, cultural, scientific, and industrial circles, as well as government employees, took part in the parade. Students from other parts of the country also participated. The overwhelming majority of demonstrators' slogans were directed against the chief leader of the State Council. Other slogans included, "Withdraw the troops and lift the martial law," and "Safeguard the constitution and guarantee human rights" (*China Daily*, 24 May 1989).

- The Association of Intellectuals was established and called upon all intellectuals to participate in the demonstration against martial law.
- Many other associations representing various professions and ordinary citizens were also established. Beijing residents were elated about their success in stopping the army.
- An overarching organization, the Joint Federation of All Circles in Beijing, was established. It combined all of the political forces emerging in the movement and was dominated by intellectuals and students.
- The Red Cross Society of China called for students to withdraw from Tiananmen Square to avoid an epidemic, and urged the government to keep its promise not to punish the students after the event.
- For four evenings in a row, local residents spontaneously gathered at all the major entrances into the city to form barricades and build roadblocks in a bid to prevent military troops from entering the city.
- Most of the People's Liberation Army troops, which had been summoned to Beijing to enforce martial law but had been blocked on the outskirts of the capital for the previous four days, had withdrawn to a temporary camp in Western Beijing. No military trucks were to be seen in southwest Beijing, for example, where local people had blocked military trucks for two days.

24 May

- Thousands of students continued their sit-in at Tiananmen Square. The General Headquarters of Tiananmen Square was established to manage the students.
- The Beijing Workers' Autonomous Union was also established. It was modeled on Solidarity, in Poland, and paralleled AUBUC. This autonomous union was a direct challenge to the official All-China Federation of Trade Unions.
- All bus and subway traffic in Beijing had been restored and social order in the city had returned to normal.
- The army troops transferred to enforce martial law in parts of Beijing were still impeded in the suburbs by masses of students and residents. Some of the troops had retreated, and it appeared that the threat of military suppression was over. But martial law was not rescinded, and the soldiers did not leave the municipality of Beijing.

25 May

- At an afternoon meeting, Chinese Premier Li Peng briefed three new foreign ambassadors to China on the situation. He said that martial law had been imposed in Beijing for six days but that the troops had not yet reached the downtown area because of blockades. He also said that anyone with common sense could see that this was not because the troops were unable to enter the downtown area, but because the government was the people's government and the People's Liberation Army was the people's own army. Because the people had not yet fully understood the meaning of martial law, the troops had exercised great restraint to avoid major clashes.
- Between 20,000 and 30,000 intellectuals participated in a demonstration against marital law organized by the Association of Intellectuals, which many of Beijing's residents also joined.
- The State Council issued an urgent circular, ordering students to be stopped from coming to Beijing. But a source within the ministry of railways expressed doubt about the effectiveness of the circular.

26 May

- It was hot, with temperatures of 96–97 degrees Fahrenheit. No one could stand the bright sunlight in Tiananmen Square, and the students' peaceful sit-in strike moved into the shadows nearby. Many students were extremely exhausted and were returning to their campuses. Some of them slept during the day and came back to Tiananmen Square in the evening.
- There were no more demonstrations in the streets.
- By 26 May, about 100,000 students had left Beijing by train to return to their colleges and universities. Although thousands of students from other parts of the country continued to flock to Beijing, a source in the railway ministry said that those students leaving the capital outnumbered those coming in.
- According to *Liberation Army Daily*, soldiers and officers of the People's Liberation Army, who were transferred to Beijing to impose martial law, said that they believed they would win the understanding of the local residents. "We are the people's army and we will serve the people just as we did in the past. How could we shoot our own people?" an officer told the paper's reporter (*China Daily*, 27 May 1989).

27 May

- At a meeting organized by the Joint Federation of All Circles in Beijing, all leaders voted that the students should leave the square. They planned to hold a large-scale demonstration to celebrate victory of the student movement for democracy, and then withdraw from the square on 30 May. However, student leaders from the General Headquarters of Tiananmen Square soon changed their minds. These student leaders were supported mainly by non-Beijing college students who had arrived in Beijing only a few days earlier and were eager to see some action. It was reported that the students wanted to stay in the square until the standing meeting of the National People's Congress held its scheduled meeting on 20 June.
- While martial law had been imposed in Beijing for eight days, most students and Beijing residents truly believed that the People's Liberation Army would never turn its guns on the people.
- Beijing officials and medical personnel again urged students to leave the "badly polluted" Tiananmen Square.
- At a meeting, a top state leader, Chairman of the Chinese People's Political Consultative Conference, said that Chinese students should realize that their current campaign for democracy had been "taken advantage of by a handful of people to create chaos," violating the original "well-intended motives" of the students themselves (*China Daily*, 28 May 1989).

28 May

- International Chinese People's Demonstration Day was on 28 May, and 20,000 students, intellectuals, and workers took to the streets to demand the resignation of Premier Li Peng and the lifting of martial law. The number of participants in the demonstration was, however, much smaller than in earlier demonstrations.
- Tens of thousands of Beijing and provincial college students still stayed in Tiananmen Square.
- Other than a traffic jam caused by the demonstrators, life was normal and orderly in most parts of the city, with martial law still being the main topic of conversation.

29 May

- Students' opinions differed as to whether to continue the sit-in or call it off.

- In a letter delivered to the university's top administrators, eight professors and associate professors from Peking University called on students to resume their classes.
- The working conference of the State Education Commission urged students to end the boycott. Graduation would be on schedule and college entrance exams would be held as planned, the official said.
- A crowd of nearly 100,000 students and workers in Tiananmen Square cheered the arrival of a 27-foot high mock Statue of Liberty made of plastic by students of the Central Art Academy of Beijing.

30 May

- In the morning, the Statue of Liberty was erected in the center of Tiananmen Square and an unveiling ceremony was held at noon. The statue drew more than 10,000 people to the square and many foreign reporters took pictures.
- The Statue of Liberty insulted the regime and disturbed some people, who did not approve of this foreign-seeming thing.
- China Central Television and Central Radio evening news broadcast a statement by the Beijing Tiananmen Square Administration demanding the immediate removal of the "illegal statue," which, it said, was "an insult to this solemn site of national celebration and state ceremonies" (*China Daily*, 31 May 1989).
- The statue became a rallying point, especially for students from universities outside Beijing, while most Beijing students felt that their major goals had been attained and that it was time to pursue a long-term strategy. Students from outside Beijing voted down the motion to evacuate the square.
- Railway transport throughout China returned to normal. Most of the 200,000 provincial students who came to Beijing to support the student strikers in the capital had gone home.
- Officers and soldiers ordered to enforce martial law received continuous instructions on the policy of "loving the capital, citizens and students," and were still stationed on the outskirts of Beijing (*China Daily*, 31 May 1989).

31 May

- About 4,000 students marched on Tiananmen Square to demonstrate against martial law.

- The number of students who stayed at the square overnight had dropped from between 100,000 and 200,000 during mid-May to around 10,000 by the end of May. Although the majority of students who continued to demonstrate at the square were from outside Beijing, a huge number of Beijing students and residents still visited daily, swelling the square's daytime population.
- Contrary to their earlier position, most student leaders decided to stay in the square until 20 June, when the standing committee of the National People's Congress was officially scheduled to convene for a regular session.

1 June

- The first day in June was International Children's Day, and parents took their children to Tiananmen Square, where families usually gathered to celebrate the holiday.
- The students' sit-in strike continued at Tiananmen Square.

2 June

- A new hunger strike was started at Tiananmen Square by four intellectuals, including a famous pop music star from Taiwan. They stayed at the Monument to the People's Heroes—the center of the square—and drew the attention of a large crowd. More than 10,000 people gathered around them in support.
- Late at night, a large Mitsubishi jeep belonging to the People's Armed Police crashed into four people in the western part of Beijing, and three of them died. The jeep's license plate had been removed to ensure that the vehicle would not be stopped by students and residents of Beijing. Bystanders converged angrily on the jeep and the crowd smashed the vehicle. News of the incident spread throughout the city. People were skeptical about the idea that the jeep was a harbinger of military invasion, and the incident galvanized students, who mobilized and paraded through the streets.

3 June

- Early in the morning, students and residents of Beijing blocked several buses that were carrying machine guns, assault rifles, and other weapons and military equipment. The blockage took place at

Liubukou, a big intersection to the west of Tiananmen Square, and it was the first time that students and residents of Beijing witnessed the fact that military equipment was actually being transferred into the city since martial law had been implemented on 20 May. Bystanders were stunned and became an angry crowd. Some students climbed atop the buses carrying weapons, and exhibited the machine guns as proof of the troops' presence in the city. At 2:00 p.m., government loudspeakers at Liubukou asked the crowd, which had grown to tens of thousands, to release the vehicles carrying arms, return any weapons that had been removed, and disperse immediately—a signal that a crackdown was finally about to begin. The crowd's immediate reaction was anger rather than fear. Shortly after the announcement, over 1,000 soldiers and armed police rushed onto the streets, firing tear gas at the protestors and clubbing anyone who stood in the way.

- All along Changan Boulevard—the main street of Beijing—students, workers, and local residents emerged to block nearby intersections and prevent troops from reaching the square. Protesters barricaded the military vehicles and let the air out of tires. The roadblocks reappeared after more than a week, during which time the streets had been clear.
- The Joint Federation of All Circles in Beijing held a meeting to discuss the situation. Although most leaders realized that it was getting serious, none believed that the government would use force to suppress the movement.
- The headquarters of the martial law troops of the People's Liberation Army and the municipal government of Beijing issued three public notices and an announcement in the afternoon. Following are excerpts:

> In the past days, a handful of people have spread rumors, viciously defaming and attacking the troops. They are also sowing dissension between the troops and the people with very dirty tricks, instigating some people to intercept and smash the military vehicles, snatching arms, beating the officers and soldiers, interrogating and examining soldiers, impeding the action of troops to impose martial law and deliberately creating disturbances to expand the turmoil.
>
> In the early morning of June 3rd, some of the martial law enforcement troops of the People's Liberation Army were ordered to march into the city to protect important objects. Due to the

> rumor-mongering and instigation of a very small number of people, some people who did not know the truth were incited to set up road blocks at many crossroads in the city, preventing some of the troops from entering the city.
>
> The People's Liberation Army troops must carry out the martial law tasks as planned and no one should prevent them. Should they be prevented, the martial law enforcement troops will take various self-defensive measures and all means to remove the resistance.
>
> The situation in Beijing at present is very serious. More serious riots can occur at any time. From now on, please do not come onto the streets and do not go to Tiananmen Square. The masses of the workers and staff members should stand fast at your posts. The residents should stay at home to secure the safety of your life and avoid unnecessary losses.
>
> (*China Daily*, 5 June 1989)

- There were ominous signs that a military crackdown was taking shape. At 6:30 p.m., government-controlled radio and television began broadcasting an announcement, throughout the city and the evening, warning people not to go out on the streets or "they should bear responsibility for their own fates" (*China Daily*, 4 June 1989).
- A new university, the Democracy University at Tiananmen Square, was established by students. An opening ceremony was held at 10:00 p.m. at the square, and the president of the university announced at the ceremony that the first day of school would be on 4 June. Many famous intellectuals would be invited to teach courses on topics such as freedom, democracy, law and order, and human rights. A large crowd was overflowing with enthusiasm.
- By the time the emergency announcement was echoing over Tiananmen Square, the army offensive against Beijing was already well underway. Changan Boulevard served as the main access route for troops heading to Tiananmen Square, and the first conflict between the army and the movement occurred at Muxudi, the western part of Changan Boulevard, a few miles away from the square.
- At Muxudi, hundreds of thousands of people, including pregnant women and teenage girls, flouted government orders and gathered to block the troops, just as they had been doing since martial law was declared on 20 May. None of them believed that the People's Army

would really fire on the people. At about 10:30 p.m., troops arrived at Muxudi armed with tanks, assault rifles, and machine guns. Unlike previous encounters, however, the troops did not stop in front of the crowd. Instead, they continually pushed their way farther east. Citizens began to throw stones and bottles at the soldiers, and then gunfire began. People could not grasp that the military crackdown was finally taking place. Most of them thought that the soldiers were simply driving people off the streets by using rubber bullets. But they soon realized that the soldiers were using live ammunition and shooting to kill. Some people were shot dead and others were wounded by gunshots.

- While blood covered the street at Muxudi, the opening ceremony of the Democracy University at Tiananmen Square was still in progress. There were more than 100,000 people at the square, and most of them sensed that the government might be planning to use some new methods to enforce martial law, but they also assumed that none of these methods would work because nothing dire had happened since martial law was implemented on 20 May. Finally, the students at Tiananmen Square heard gunshots, and people from Muxudi and other parts of the city rushed into the General Headquarters of Tiananmen Square to report that the army had killed students and residents. Students at Tiananmen Square were shocked. They were also extremely angry, which drove away their fear. While some people left the square, all of the students began to gather at the Monument to the People's Heroes, the center of the square, under the order of the General Headquarters of Tiananmen Square.
- When the troops headed to Tiananmen Square from the east, west, and south, unarmed Beijing residents took to the streets to block their entry and protect the students in the square. The troops opened fire and a row of three-wheeled flatbed bicycles began taking wounded people out of the crowds. Some of the wounded people died on the way to hospitals. In the emergency rooms, people were lain down on the floors with bullet wounds to the stomach, chest, and head.

4 June

- At the General Headquarters of Tiananmen Square, student leaders were divided about what they should do. Some of them insisted that the students stay at Tiananmen Square no matter what happened, while others believed that they should withdraw. Some, who wanted them to stay, held fast to the principle of peace, reason, and non-violence, while others asked the unarmed crowd to resist the military crackdown. At about 1:00 a.m., the General Headquarters of Tiananmen Square finally decided that the students should stay and hold onto the principle of peace, reason, and non-violence.

- Thousands of students at Tiananmen Square made an oath together, pledging to devote their lives to democracy and their motherland, etc.
- Troops finally arrived at Tiananmen Square at about 1:30 a.m. and began to encircle the square. The army's objective at that point was not to take the square, but to clear everyone off Changan Boulevard at the north end of the square and establish a cordon to prevent anyone outside the square from joining those already there. Many people on the West and East Changan Boulevard tried to break down the circle—and every five or ten minutes a burst of gunfire was directed toward the crowds. People dispersed, dived for cover, and ran for the trees along the side of the street, but then quickly regrouped and went back into the street. After every burst of gunfire, several people went down. They either died or were seriously wounded.
- Everyone worried about what would happen to the thousands of students inside the cordon at Tiananmen Square.
- While more and more troops arrived at Tiananmen Square at 3:00 a.m., the four intellectuals who had begun their hunger strike there on 2 June decided to end their hunger strike immediately and

  1 persuade student leaders to give up their decision to stay in the square,
  2 ask student leaders to take the thousands of students out of the square, and
  3 negotiate with the army for the students' peaceful withdrawal.

- At 3:30 a.m., two intellectuals served as representatives at a meeting with one of the military leaders at Tiananmen Square. The army accepted their demand for the students' peaceful withdrawal.
- At 4:00 a.m., all lights in Tiananmen Square suddenly went off and deep darkness covered the square. Troops with tanks could be heard beginning to move around the square.
- At 4:30 a.m., all lights in the square came back on and the students began to withdraw from Tiananmen Square.

3 Anti-threat resistance: a game with incomplete information

As discussed in Chapter 1, a newly created social system with certain properties is the singular condition that is both necessary and sufficient to explain the tragic events at Tiananmen Square in 1989. The emergence of social movements generates a new social system composed of the movement and the state, and the actions of movement participants interface with those of the state to produce new properties of the system, which lead to outcomes and consequences for the social movements themselves. One of these new properties involves the effectiveness of state-issued threats. Will such threats hinder social movements? Under what conditions will these threats fail, and why do unarmed movement participants often confront armed police officers?

In 1989, one of the Chinese government's central strategies in response to the emergence of the student movement was the use of threats to compel movement participants to cease their activities. The first threat was issued on 26 April 1989, in an official editorial that characterized the student movement as a "planned conspiracy" and "turmoil," the essence of which was "basically to repudiate the leadership of the Chinese Communist Party and to repudiate the socialist institutions" (*People's Daily*, 26 April 1989). The editorial favored immediate and harsh suppression of the movement. Following this threat, thousands of armed police officers were deployed to Tiananmen Square from a western suburb of Beijing, where most of the universities were located.

The students were unarmed and did not know what the police were going to do, but they decided not to back down. More than 30,000 students had marched through the streets and confronted the police face to face. The armed police officers backed down—the threat therefore failed in its objective to stop the protests. The total number of participants increased from 60,000, before the show of police force, to 250,000 afterward. Why did the threats issued by the state fail to hinder the social movement?

Preference structure: the mechanics of interaction

According to Morris and Herring (1987), high levels of repression and low levels of power can force challengers to withdraw from collective action. Zald introduced the internal strategic explanation that "the state generates many of the issues with which social movements wrestle; as well, the state facilitates or hinders movements, lowering or raising the costs of collective action, operation in coalition with the movements or opposing it" (1992:339). The core postulate here is that the state plays an important role in determining a movement's development and outcomes, and high levels of suppression and limited access to power can force challengers to withdraw from collective action. In reality, however, this is not always the case. There is much evidence which shows that challengers who have no access to power may nevertheless confront a powerful state. They choose to "resist" after the state's threats. Why is there inconsistency between the core postulate and the reality?

While many scholars of social movements recognize that the interaction between movements and states is the central element in a movement's development (Tilly, 1978; McAdam, 1982, 1983; Grant II and Wallace, 1991; Quadagno, 1992; Morris, 1983; Koopmans, 1993), no one has ever discussed the preference structures connecting the movement and the state. This omission is due to the absence of appropriate concepts, language, and techniques.

Preference structure is a core element of the newly created social system. It is what induces and constrains both the movement and the state vis-à-vis their decision-making process. Kreps regards preference structures as "the mechanics of the interaction—who does what when with what information" (1990b:87). Coleman describes preference structures as "the rules of a game so that reasonable persons, as players, would act so as to generate system behavior" (1994:33). Morrow (1994) believes that the rules for reaching group decisions, the outcomes that result from actors' choices in a game, and the choices that people see or do not see in a situation are all structural. The preference structure is an interaction framework that clearly displays the fact that both sides' actions are determined, not only by their own goals, but also by the actions of others.

Scholars have traditionally studied the interaction between movements and the state in an individual framework where the mechanics of interaction—or preference structures—do not exist. The individual framework is therefore not appropriate to characterize the outcomes of social movements that are systemic consequences.

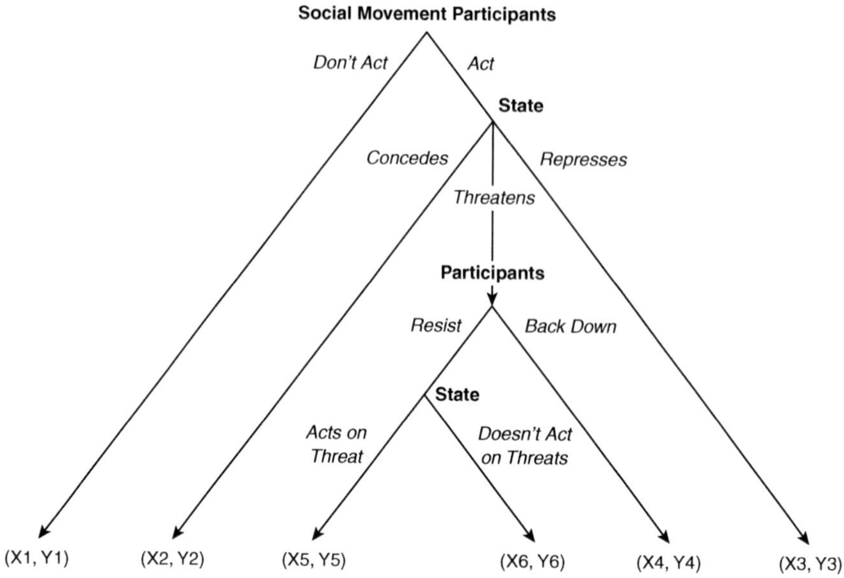

Figure 3.1 A model of movement development

A model of movement development

In studying preference structures connecting movements and the state, the key is to draw an extensive-form game that models the process of movement development (see Figure 3.1).

In Figure 3.1, the main actors are social movement participants and the state. The model specifies the sequence of choices made by both, the alternatives from which each party must choose at each juncture, and the information each has when making choices based upon the decisions made by the other side. Within every parenthesis, the first component is the participants' payoff, and the second is the state's payoff.

While the movement is developing, confrontation is sequential in nature: people in the movement take demonstrative actions, the state responds, the participants reply again, and this process continues until one side backs down. The model of movement development presented here is a four-step sequence, although the sequence might be repeated many times. While the model is much simpler than in any particular case, it captures the key structural features of players' interactions.

In the first step, a social movement begins with the participants' decision to demonstrate. In Figure 3.1, then, the game is initiated by a simple choice between two alternatives: either the participants demonstrate, or they do not.

In the second step, following the participants' decision to act, it is the state's turn to choose. After the emergence of a social movement, the state usually has a variety of options. For the present, however, it will serve our discussion to reduce them to just three:

1 the state can make a concession to participants by accepting their requirement, thus ending the game;
2 the state can follow a strategy of repression and force movement participants to withdraw; or
3 the state can attempt to induce participants to back down, thereby continuing the game. The only way to induce participants to back down is to threaten them. If they do not back down, the state makes it clear that the consequences will be worse than if they did.

In the third step, participants have an opportunity to respond to the state's threat. The response might be "confrontational actions that are also non-violent, but they aim to disrupt official policies or institutions, and for that reason are usually illegal" (Koopmans, 1993:640). In Figure 3.1, confrontational actions have been denoted by "Resist." Another response might be to back down.

In the fourth step, the state chooses whether to act on its initial threat to use force after observing the participants' response. If state officials choose this option, movement participants are forced to withdraw and the game ends. If the state chooses not to use force, the game may repeat or end when the state accepts the movement's requirement.

This represents a dynamic game with perfect information. The payoffs for both players are public knowledge, and individual players can consistently rank their payoffs in order of preference—a ranking called the "preference function." In the game, preference functions for both players are also public knowledge. This knowledge derives from players' common sense thinking about a situation, based on an assumption that people have conventional modes of behavior.

Economists describe the process of choosing the best decision as the "maximization of utility." Utility is the variable whose relative magnitude determines the direction of preference. According to Hirshleifer and Glazer:

> Most modern economists do not believe that utility can be given a cardinal measure. If people can state that they prefer two million to one million—but not by how much—their utility is said to be an ordinal magnitude.
>
> (1992:63)

Table 3.1 Payoffs and ordinal utility for both sides

Strategies	*Movement's payoffs*	*State's payoffs*	*Movement's utility*	*State's utility*
Participants don't act	X1 = 0	Y1 = 0	U (X1) = 4	U (Y1) = 4
Participants act, state concedes	X2 = G	Y2 = –P	U (X2) = 5	U (Y2) = 1
Participants act, state represses	X3 = –D	Y3 = –E	U (X3) = 1	U (Y3) = 3
Participants act, state threatens, participants back down	X4 = –L	Y4 = B	U (X4) = 3	U (Y4) = 6
Participants act, state threatens, participants resist, state uses force	X5 = –D + H	$Y5 = C - \sqrt{E}$	U (X5) = 2	U (Y5) = 5
Participants act, state threatens, participants resist, state doesn't use force	X6 = G + H	$Y6 = -P - C + E^2$	U (X6) = 6	U (Y6) = 2

Table 3.1 shows the payoffs and the ordinal utility for both movement and state within the framework of the game.

In Table 3.1, the participants' payoffs are as follows:

- X1 = 0 (status quo: payoffs are normalized to 0)
- X2 = G (G denotes a substantial gain for the participants, such as human rights improvement or legitimization of civil organizations, etc., when the state makes a concession or backs down. $G > 0$)
- X3 = –D (–D denotes huge loss, including death for the participants. $D > 0$, and $-D < 0$)
- X4 = –L (the participants lose face when they back down. Consider it as –L, a general loss that is smaller than the huge loss for participants. $0 < L < D$, $-L < 0$)
- X5 = –D + H (when the state uses force, the loss for participants is –D, a huge loss. H denotes honor to the participants who have resisted the threat before the state uses force. $G > H \geq L$)
- X6 = G + H (the participants resist and the state backs down, which brings both substantial gain and also honor to the participants).

In Table 3.1, U denotes utility. The participants' ordinal preferences can be represented by a descending sequence of numbers. The largest number is assigned to the most preferred outcome, the second largest number to the next outcome in the preference order, and so on, down to the least preferred outcome. The participants' ordinal utility is as follows:

- U (X6) = 6 is the best outcome for participants, who get substantial gain and honor
- U (X2) = 5 is the second best. Participants get substantial gain without getting honor
- U (X1) = 4 involves neither gain nor loss for the participants
- U (X4) = 3 means that the participants lose face because they have given up
- U (X5) = 2 means that some participants would die with great honor
- U (X3) = 1 is the worst result for the social movement, and some participants will die.

In Table 3.1, the state's payoffs are as follows:

- Y1 = 0 (status quo: payoffs are normalized to 0)
- Y2 = –P (P denotes some power or authority that the state relinquishes when it concedes. P > 0)
- Y3 = –E (E denotes legitimacy that the state loses when it represses a social movement. Assume that the state would rather lose legitimacy than power. P > E > 0)
- Y4 = B (B denotes big gains for the state after the participants withdraw, because the state not only keeps its power intact, but also shows that its authority is unchallenged. B > P)
- Y5 = C $-\sqrt{E}$ (C denotes credibility that the state earns by keeping its word to use force. Although the state loses legitimacy by repressing, its loss is smaller than E because it issued a threat before implementing repression, thereby giving participants an opportunity to withdraw. Therefore, E becomes $\sqrt{E}$. Assume C = E, when E > $\sqrt{E}$, C > $\sqrt{E}$, and [C – $\sqrt{E}$] > 0)
- Y6 = –P – C+ E^2 (the state loses power or authority—and credibility—by not using force, but gains legitimacy two times by not repressing the movement in the second step and the last step. E^2 > C).

In Table 3.1, the state's ordinal utility is as follows:

- U (Y4) = 6 is the best outcome for the state. The movement withdraws without use of state force
- U (Y5) = 5 is the second best. The state loses no power, and gains credibility, but loses some legitimacy
- U (Y1) = 4 means neither gain nor loss for the state

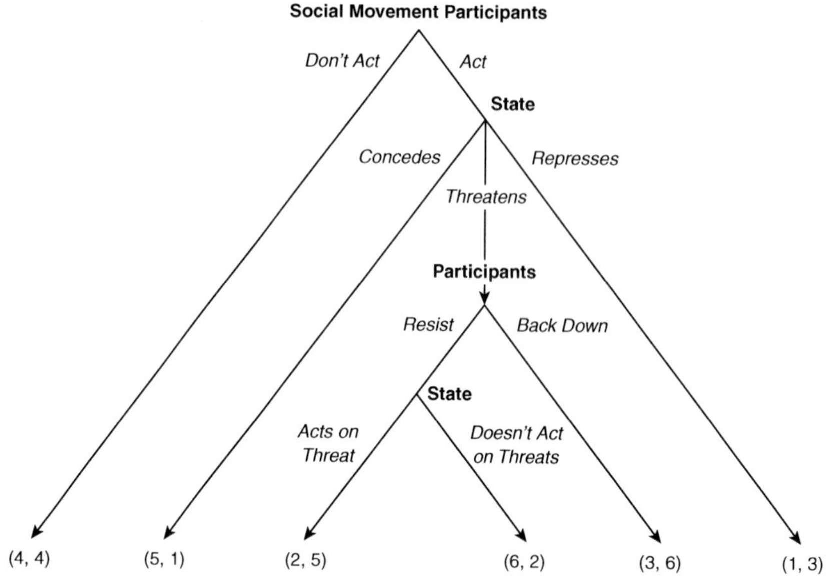

Figure 3.2 A model of movement development with ordinal utility

- U (Y3) = 3 means a loss of legitimacy for the state
- U (Y6) = 2 means some loss of power and credibility for the state, but a big gain in legitimacy
- U (Y2) = 1 is the worst outcome for the state, which relinquishes some power or authority.

Figure 3.2 presents a model of movement development as a four-step sequence or a preference structure connecting the movement participants and the state. The difference between Figures 3.1 and 3.2 is that Figure 3.2 shows the players' utility measured by an ordinal magnitude.

What is a solution concept in the game in Figure 3.2? In game theory, a solution concept is defined as a prediction about the choices of players and the corresponding outcomes in a game. The game in Figure 3.2 can be solved by backward induction. Starting at the fourth step, the state faces a choice between payoff 5 (acting on its initial threat to use force) and payoff 2 (not acting on its threat—not using force), and the state chooses to act on its initial threat. At the third step, the participants anticipate that the state will use force at the fourth step, which yields payoff 2 for the participants. Therefore, the choice for the participants at the third step is between payoff 2 (resisting) and payoff 3 (backing down); backing down is the optimal decision.

At the second step, the state knows that the participants will back down when the game reaches the third step, which yields payoff 6 for the state. The choice for the state at the second step is between payoff 1 (conceding), payoff 3 (repressing), and payoff 6 (threatening), so threatening is optimal. At the first step, participants anticipate that the state will choose to use force at the last step, and that they will be forced to back down at the third step, so the state will threaten participants at the second step, which yields payoff 3 for the participants. The choice for participants at the first step is between payoff 4 (not taking action) and payoff 3 (taking action); inaction is optimal.

The backward-induction outcome of this game includes several possibilities. The first possibility is that movement participants choose not to take action in the first step, thereby ending (or precluding) the game. The second possibility is that the game does not end at the first step because participants think that the state may not act rationally. They may therefore choose to take action at the first step, hoping that the state will make concessions at the second step. The backward-induction outcome tells us what happens if the game does not end at the first step.

According to the backward-induction outcome, the game should end either at the first step, where the participants choose not to take action, or at the third step, where they back down after the state threatens them. This game theoretical prediction about movement development and outcomes is consistent with the internal strategic explanation, which means that the state plays an important role in determining a movement's development and outcomes, based on power contrast, and that high levels of suppression and limited access to power can force challengers to withdraw from collective action.

Private information and a changed preference structure

Although backward induction is an appealing way to solve games with complete information, inconsistencies occur between theoretical prediction and empirical evidence. During the 1989 Chinese student movement, for example, the state, backed by strong military force, threatened the movement three times. According to the model in Figure 3.2, the unarmed students and other participants should have backed down in the face of these threats. In sharp contrast to the prediction, however, the students and other participants never backed down—they always chose to resist after the threats were made. Why are the theoretical prediction and the empirical evidence inconsistent?

In order to answer this question, it is necessary to discuss whether players' preferences are always common knowledge. In other words, it is necessary to specify what information about the players' preferences is commonly known, and what information about those preferences is private. According to Fearon, "A person who knows something that others do not know about his preferences, intentions, or any other variable is said to have private information" (1992:113).

In the model (Figure 3.2), it is certain that the preferences of movement participants are common knowledge because any substantial gains are better payoffs than honor alone, and both gains and honor are better than the status quo. For the participants, being repressed under any circumstances is worse than losing face. They lose face when they back down after the state threatens them, but being repressed after resisting definitely looks better than being repressed when they have not resisted.

In contrast to the participants' preferences, the state's preferences are not commonly known to the same extent, and this uncertainty is linked to the issue of private information. In Figure 3.2, for example, it has been assumed that the state would rather lose legitimacy than lose any power or authority (P > E > 0), which is common knowledge. But is E always smaller than P? In other words, does the state always prefer losing legitimacy to losing any power? Movement participants cannot answer this question under certain circumstances because they do not have the necessary information.

Considerable empirical evidence suggests that a state sometimes prefers losing legitimacy to losing authority, which means that its agents adhere to their initial threat to use force (i.e., P > E > 0). In her analysis of the emergence of terrorism in West Germany, for example, Karstedt-Henke (1980) argues that authorities overreacted to the emergence of protest in the initial phase of mobilization, and, in an attempt to quell unrest, followed a strategy of repression. But because they were caught off guard, they did so in an inconsistent and undifferentiated way that provoked public outrage and led to further protests. However, the state sometimes prefers acquiring or maintaining an image of legitimacy over keeping its power or authority intact, which means that the state does not act on its initial threat or use force after participants resist (i.e., E > P > 0).

What determines a state's preferences? Why does the state hold its preferences as private information? Chapter 4 will answer these questions, but it is necessary to specify here that it is possible that movement participants are uncertain about what the state will do at the step that follows the threats it issues. Movement participants can only draw

Table 3.2 Unchanged payoffs and changed ordinal utility

Strategies	*Movement's payoffs*	*State's payoffs*	*Movement's utility*	*State's utility*
Participants don't act	X1 = 0	Y1 = 0	U (X1) = 4	U (Y1) = 5
Participants act, state concedes	X2 = G	Y2 = –P	U (X2) = 5	U (Y2) = 3
Participants act, state represses	X3 = –D	Y3 = –E	U (X3) = 1	U (Y3) = 1
Participants act, state threatens, participants back down	X4 = –L	Y4 = B	U (X4) = 3	U (Y4) = 6
Participants act, state threatens, participants resist, state uses force	X5 = –D + H	Y5 = C – $\sqrt{E}$	U (X5) = 2	U (Y5) = 2
Participants act, state threatens, participants resist, state doesn't use force	X6 = G + H	Y6 = –P –C + E^2	U (X6) = 6	U (Y6) = 4

inferences about private government information by examining the state's words and deeds.

How does the private information held by the state influence movement development? This information can totally change the movement's process of development and its outcomes, because choices made by the players in light of their own preferences and their expectations about the preferences of others determine the solution for the game. When the preferences of either side change, the choices made by both sides will change. As a result, the movement's development and outcomes can change significantly.

Table 3.1, for example, is based on the assumption that the state prefers losing legitimacy to losing any power or authority, which means that it sticks to its initial threat to use force (i.e., P > E > 0). But, in Table 3.2, the state prefers its legitimacy to keeping its power intact (i.e., E > P > 0). Also, in Table 3.2, it is assumed that the state's agents believe that legitimacy is more important than credibility (i.e., E > P > C > 0, C < $\sqrt{E}$, and [C–$\sqrt{E}$] < 0). In Table 3.2, there are no changes in the payoffs for the two sides, but there are marked changes in the state's preferences, while the participants' preferences remain unchanged (see Table 3.2).

In Table 3.1, U (Y4) > U (Y5) > U (Y1) > U (Y3) > U (Y6) > U (Y2).

In Table 3.2, U (Y4) > U (Y1) > U (Y6) > U (Y2) > U (Y5) > U (Y3). The state's ordinal utility is as follows:

- U (Y4) = 6 is the best outcome for the state. The movement withdraws and the state does not use force
- U (Y1) = 5 is the second best outcome. There is neither gain nor loss for the state
- U (Y6) = 4 means that the state loses some power and credibility, but gains a lot of legitimacy
- U (Y2) = 3 means that the state loses some authority, and gains nothing
- U (Y5) = 2 means that the state loses some legitimacy, but retains credibility
- U (Y3) = 1 is the worst outcome for the state: it loses legitimacy.

A model of movement development with changed ordinal utility is presented below in Figure 3.3.

According to backward induction, the state faces a choice starting at the fourth step. This choice is between payoff 2 (acting on the initial threat and using force) and payoff 4 (not acting on the threat—not using force), so the state chooses not to act on its initial threat. At the third step, the participants anticipate that the state will not use force at the fourth step, which yields payoff 6 for the participants. Therefore, the

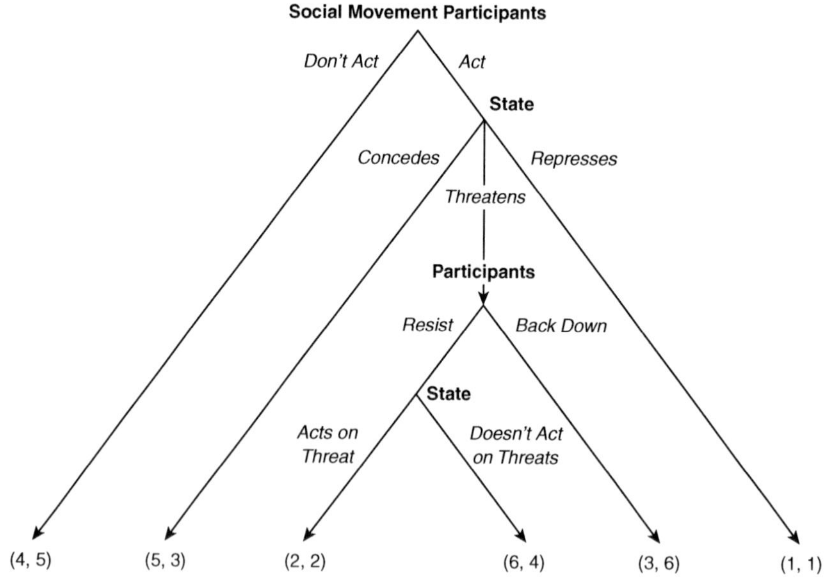

Figure 3.3 A model of movement development with changed ordinal utility

choice for participants at the third step is between payoff 6 (resisting) and payoff 3 (backing down), so resisting is the optimal choice.

At the second step, the state knows that the participants will resist when the game reaches the third step, which yields payoff 4 for the state. The choice for the state at the second step is between payoff 3 (conceding), payoff 1 (repressing), and payoff 4 (threatening), so threatening is optimal. At the first step, the participants anticipate that the state will choose not to use force at the last step, and they will choose "Resist" at the third step, so the state will threaten at the second step, which yields payoff 6 for the participants. The choice for the participants at the first step is between payoff 4 (not taking action) and payoff 6 (taking action), so taking action is optimal.

In game theory, Nash equilibrium is the combination of strategies chosen by players in which no player could do better by choosing a different strategy, given the strategy the other chooses. In the game in Figure 3.3, Nash equilibrium is that the movement participants choose to act—then the state threatens and the participants resist the threat, but the state does not act on its initial threat. The Nash equilibrium in Figure 3.3 is totally different from that in Figure 3.2, because the backward-induction outcome in Figure 3.2 is that movement participants either choose to act or not to act, and then the state threatens and the participants back down. The reason for this difference is that, in Figure 3.2, the state prefers to keep its authority intact rather than retain or acquire legitimacy, and in Figure 3.3 its preferences have reversed.

In short, state leaders keep their preferences private in most cases. Therefore, uncertainty linked to private information held by the state is critically important for the participants of a social movement.

A game with incomplete information

Although two models with two different preference structures in Figures 3.2 and 3.3 have been discussed, neither illustrates the real process of movement development in which the state's preferences are private information, because the participants do not know whether the state prefers legitimacy to keeping its authority intact. It is therefore necessary to construct a new model that incorporates the real features of interaction between a movement and a state.

This new model differs from Figures 3.2 and 3.3 in the following three ways. First, the models in Figures 3.2 and 3.3 are games with complete information (i.e., games in which the players' payoffs and preferences are common knowledge), although the players' payoffs and preferences in the two games are different. The new model, on the other hand, is a game with incomplete information (i.e., a game in which at least one player is uncertain about another player's preferences). In this new model, the movement is not confronted by a state that prefers to keep its authority intact. Nor is it confronted by one that prefers legitimacy, because movement participants simply do not know what the state's preferences are.

Second, although backward induction is a very neat way to solve the games in Figures 3.2 and 3.3, it cannot be used in the new model because it can only predict outcomes of games with complete information, and the new model is a game with incomplete information.

Because the new model is constructed on a foundation of imperfect knowledge, it is necessary to introduce the concept of perfect Bayesian equilibrium, another solution concept, to solve the game in the new model. As Gibbons wrote:

> The crucial new feature of this equilibrium [perfect Bayesian equilibrium] concept is due to Kreps and Wilson (1982): beliefs are elevated to the level of importance of strategies in the definition of equilibrium. Formally, equilibrium no longer consists of just a strategy for each player but now also includes a belief for each player at each information set at which the player has the move. The advantage of making the players' beliefs explicit in this way is that, just as in earlier chapters we insisted that the players choose credible strategies, we can now also insist that they hold reasonable beliefs both on the equilibrium path and off the equilibrium path. (1992:179)

Since then, perfect Bayesian equilibrium—the solution concept in the game with incomplete information—has been applied to many problems in economics and political science. For example, Fearon (1992) argues that international crisis bargaining occurs because state leaders typically have private information about their willingness to use force rather than accept possible negotiated settlements, and because they may have incentives to misrepresent this private information.

Third, both Figures 3.2 and 3.3 present extensive-form games with a four-step sequence that models a process of movement development. The new model, however, presents a threat subgame in the extensive-form game. This threat subgame begins at one of the state's decision nodes at the second step of the extensive-form game. And it is the threat subgame that must be analyzed; other parts of the extensive-form game can be ignored for the time being. The new model is presented in Figure 3.4, and, unlike the earlier models, the first component presented here within every parenthesis represents the state's payoff, and the second one represents the participants' payoff because, in this subgame, the state is the player who moves first, and the participants are the second player.

The threat subgame in Figure 3.4 is a game with incomplete information because the participants do not know the state's preferences. By definition, the state is tough if it prefers retention of its power/authority to legitimacy (i.e., if the state uses force after participants resist its threat). The state is humane if it prefers legitimacy to keeping power (i.e., the state does not use force).

In the first step, p represents the probability that the participants believe the state is tough, and 1–p represents the probability that the participants believe the state is humane. The game is initiated by a simple choice between two alternatives made respectively by both the tough state and the humane state: to show or not to show force after issuing a threat. If the state chooses not to show force, the threat fails, the participants succeed, and the game ends.

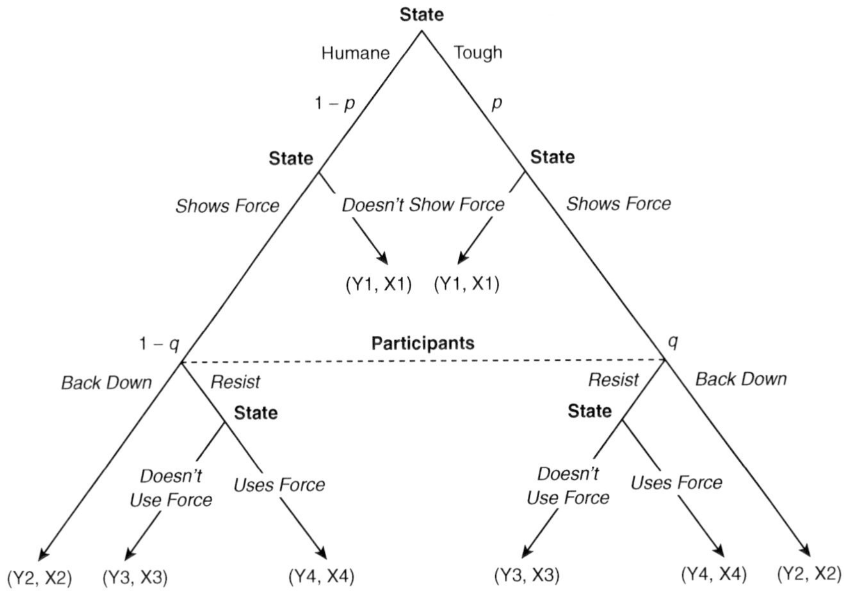

Figure 3.4 A threat subgame: a game with incomplete information

In the second step, the dotted line represents an information set. The participants have this information set, with two nodes that reflect their inability to verify the node or its location, and when they must make a decision. In other words, although the participants have observed the state's show of force after its threat, they do not know which type of state they confront. Suppose the participants are located on the tough-state side, with probability q, and on the humane-state side with probability 1–q. On either side, they have a choice between resisting and backing down. If the participants choose to back down, the game ends.

In the third step, the state has to choose whether or not to use force after the participants choose to resist its threat.

In Figure 3.4, the participants' preferences are common knowledge, while the state's preferences are private information. But it is known that the tough state would use force after the participants resist threats, and the humane state would not use force. The payoffs and the ordinal utility for both sides are summarized in Table 3.3.

In Table 3.3, the movement's payoffs and ordinal utility are as follows:

- X1 = G (G denotes a substantial gain for participants, such as human rights improvement or legitimization of civil organizations, etc., when the state backs down. $G > 0$)
- X2 = –L (participants lose face when they back down. Consider it as –L, a general loss. $-L < 0$)
- X3 = G + H (H denotes honor to the participants who have resisted the threat before the state uses force. The participants resist and the state backs down, which brings the participants not only substantial gain, but also honor. $G > H \geq L$)
- X4 = –D + H (–D denotes huge loss, including death, for the participants, $D > L$).

In Table 3.3, U (X3) > U (X1) > U (X2) > U (X4).

In Table 3.3, the state's payoffs and ordinal utility are as follows:

- Y1 = –P (P denotes power or authority that the state relinquishes when it backs down. $P > 0$)
- Y2 = B (B denotes big gains for the state after the participants withdraw, because the state keeps its power intact and also shows that its authority is unchallenged. $B > P$)
- Y3 = $-P - C + E^2$ (C denotes credibility, $B > C > 0$. E denotes legitimacy, $B > E > 0$. The state loses power or authority—and credibility—by

Table 3.3 Payoffs and ordinal utility in the game with incomplete information

Strategies	*Movement's payoffs*	*State's payoffs*	*Movement's utility*	*Tough state's utility*	*Humane state's utility*
State doesn't show force	$X1 = G$	$Y1 = -P$	$U(X1) = 3$	$U(Y1) = 1$	$U(Y1) = 1$
State shows force, participants back down	$X2 = -L$	$Y2 = B$	$U(X2) = 2$	$U(Y2) = 4$	$U(Y2) = 4$
State shows force, participants resist, state doesn't use force	$X3 = G + H$	$Y3 = -P - C + E^2$	$U(X3) = 4$	$U(Y3) = 2$	$U(Y3) = 3$
State shows force, participants resist, state uses force	$X4 = -D + H$	$Y4 = C - \sqrt{E}$	$U(X4) = 1$	$U(Y4) = 3$	$U(Y4) = 2$

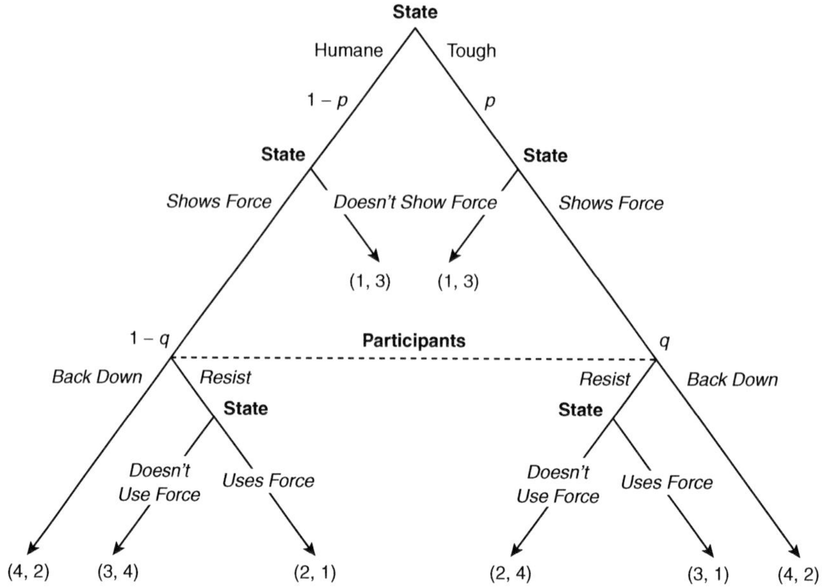

Figure 3.5 A threat subgame with ordinal utility

not using force, but gains legitimacy twice by not repressing at the beginning and end)

- Y4 = C – $\sqrt{E}$ (the state earns credibility by keeping its word to use force. Although the state lost legitimacy by repressing, the loss is smaller than E because the state first issued threats, which gave participants an opportunity to withdraw. Therefore, E becomes $\sqrt{E}$).

There is a significant difference in the preferences between the tough state and the humane state. The tough state prefers to keep power and authority intact, while the humane state prefers legitimacy. It is therefore necessary to discuss their preferences separately. In Table 3.3, the tough state's preferences are represented as U (Y2) > U (Y4) > U (Y3) > U (Y1), and the humane state's preferences are U (Y2) > U (Y3) > U (Y4) > U (Y1).

The game of incomplete information with ordinal utility for both sides is presented in Figure 3.5, and the central issue here is how to find a perfect Bayesian equilibrium of this game.

In this game, the tough state has a dominant strategy (showing and using force), because U (Y4) > U (Y3) > U (Y1) for the tough state. The humane state also has a dominant strategy (showing force but not using it), because U (Y3) > U (Y4) > U (Y1) for the humane state. For both states, the first move in their dominant strategy is the same (showing force).

Therefore, there are no separating equilibriums or semi-separating equilibriums in this game. The only case to be considered here is whether there is a pooling equilibrium, which would be a perfect Bayesian equilibrium in this game.

As mentioned before, the first move for both the tough state and the humane state in their dominant strategy is to show force. Therefore, the only case to be considered for the pooling equilibrium is the one in which both the tough and the humane states choose to show force in their first move. In this case, $q = p$. According to Bayes's rule, the state's action does not disclose any new information.

If the participants choose to back down, their expected payoff is $pX2 + (1 - p)\ X2 = X2$.

If the participants choose to resist, the tough state will use force and the humane state will not. Therefore, the participants' expected payoff is $pX4 + (1 - p)\ X3 = p(X4 - X3) + X3$.

Assume that the participants are equally inclined to their two strategies—backing down or resisting. In that case, $X2 = p(X4 - X3) + X3$.

$$p^* = \frac{X2 - X3}{X4 - X3} = \frac{X3 - X2}{X3 - X4}$$

p^* is called a critical value of p. While the participants believe that the state is tough, with probability p, there are three possible relations between p and p^*. When $p = p^*$, the participants' expected payoff from backing down would be the same as that from resisting. When $p > p^*$, $X2 > p(X4 - X3) + X3$ (X2 is the participants' expected payoff from backing down, and $p(X4 - X3) + X3$ is the participants' expected payoff from resisting), then the participants choose to back down. However, when $p < p^*$, $p(X4 - X3) + X3 > X2$, then the participants choose to resist.

The first pooling equilibrium in this game is the following:

σ: Tough state: chooses to show force and use force
Weak state: chooses to show force but not to use force
Participants: choose to back down
μ: $q = p\ (p > p^*)$

The second pooling equilibrium in this game is the following:

σ: Tough state: chooses to show force and use force
Weak state: chooses to show force, but not to use force
Participants: choose to resist
μ: $q = p\ (p < p^*)$

In these two cases, none of the players in the game deviates; therefore, the two pooling equilibriums are perfect Bayesian equilibriums.

Let A denote the probability that the participants resist the threat. Then:

$$A = 1, \quad \text{if} \quad p < p^*$$
$$A = 0, \quad \text{if} \quad p > p^* \qquad (1)$$
$$A = \frac{1}{2} \ \text{otherwise.}$$

$$p^* = \frac{X3 - X2}{X3 - X4} (X3 > X2 > X4 \ \text{ and } \ 1 > p^* > 0) \qquad (2)$$

where p is the probability that the participants believe the state is tough, p^* is a critical value of p, and X2, X3, X4 are the participants' payoffs.

In equation (1), if the participants' belief, p, that the state is tough is smaller than p^*, they resist the threat. If p is greater than p^*, the participants back down. In the game with incomplete information, the participants' initial belief p is usually 50 percent. In other words, when participants are uncertain about the state's preferences; they always believe that chances are 50/50 that the state will use force. When p = 50 percent, is p greater or smaller than p^*? In other words, if the participants are uncertain about the state's preferences after it issues threats, will they back down or will they resist?

One way to find out the answer to this question is to determine the conditions under which p^* = 50 percent.

From (2), $p^* = \dfrac{X3 - X2}{X3 - X4} (X3 > X2 > X4 \ \text{ and } \ 1 > p^* > 0)$

If $p^* = \dfrac{X3 - X2}{X3 - X4} = 0.5$

then $(X3 - X2) = 0.5(X3 - X4)$, and

$X3 - X2 = 0.5X3 - 0.5X4$, and

$(X3 - 0.5X3) - (0.5X2 + 0.5X2) = -0.5X4$, and

$0.5X3 - 0.5X2 = 0.5X2 - 0.5X4$, and

$$X3 - X2 = X2 - X4 \qquad (3)$$

Equation (3) is a condition under which $p^* = 50$ percent. According to equation (3), if $(X3 - X2) > (X2 - X4)$, then $p^* > 50$ percent, $p^* > p$ ($p = 50$ percent); the participants would choose "Resist" based on the game theoretical model. But, if $(X3 - X2) < (X2 - X4)$, then $p^* < 50$ percent, $p^* < p$ ($p = 50$ percent), and participants would choose "Back Down." What is a condition under which $(X3 - X2) > (X2 - X4)$? In other words, under what condition would participants choose "Resist," although they are uncertain about the state's preferences?

In Table 3.1, X3 is the participants' payoff when the state shows force, the participants resist, and the state does not use force; $X3 = H + G$. X2 is the participants' payoff when the state shows force and the participants back down; and $X2 = -L$. X4 is the participants' payoff when the state shows force, the participants resist, the state follows with force, and $X4 = -D + H$.

$$X3 - X2 = H + G - (-L) = H + G + L$$

$$X2 - X4 = -L - (-D + H) = D - L - H$$

Assume $(X3 - X2) > (X2 - X4)$:

$$\because (X3 - X2) > (X2 - X4)$$

$$\therefore (H + G + L) > (D - L - H)$$

$$\therefore (G + H + H) > (D - L - L)$$

$$\therefore [G + H + (H - D)] > (-L - L)$$

$$\because (H - D) < -L \text{ (see Table 3.1)}$$

$$\therefore [G + H + (H - D) - (H - D)] > [-L - L - (-L)]$$

$$\therefore (G + H) > -L \tag{4}$$

As long as $(G + H) > (-L)$, and then $(X3 - X2) > (X2 - X4)$. In Table 3.1, $(G + H)$ is the participants' payoff from resisting and not being repressed, and $(-L)$ is their payoff from backing down. In the model in Figure 3.5, $(G + H)$ is always greater than $(-L)$. Therefore, the condition under which $(X3 - X2) > (X2 - X4)$, does exist. When $(X3 - X2) > (X2 - X4)$, then $p^* > 50$ percent, $p^* > p$ ($p = 50$ percent). According to the perfect Bayesian equilibrium, the participants would resist if they believe the chance to be 50/50 or less than 50 percent that the state will use force.

Although the three models in Figures 3.2, 3.3, and 3.5 have been discussed, only the model in Figure 3.5—a game with incomplete information—has depicted the real process of movement development, because the preferences of state leaders are always private information to a certain extent, and the uncertainty linked to this private information is of critical importance for the movement's development and outcomes.

There are two important features in the model of Figure 3.5. First, the participants know nothing about the state's preferences. They do not know what will happen after they resist the state's threat. The only thing they know is that the state will use force if it prefers to keep its power and authority, and it will not do so if it prefers legitimacy. Second, the participants have their own beliefs about the state's preferences. They believe that the odds are approximately even when it comes to the question of the state using force.

The following proposition about the strategy chosen by the participants in a game with incomplete information is drawn from the model in Figure 3.5:

> *Proposition 1: in a game with incomplete information, anti-threat resistance always occurs. The movement will choose "Resist" if the participants believe that the chances are 50/50, or less than 50 percent that the state will make good on its threats, and if the movement's payoff from resisting and not being repressed is greater than that from backing down.*

Conclusion

The important finding in this chapter is that state-issued threats are usually ineffective in a movement's development. A proposition based on the game with incomplete information specifies the conditions under which a state-issued threat is likely to fail in its objectives and participants are likely to resist. Anti-threat resistance has been defined as the participants' resistance after a state threatens a movement. According to the first proposition, this anti-threat resistance is very likely to occur after the movement has emerged.

The discussion in this chapter has demonstrated the importance of studying the preference structures connecting a movement and a state. These structures change dramatically when players change their preferences. Therefore, various preference structures lead movements to develop differently, and to reach different outcomes.

Table 3.4 The effect of preference structures on movement outcomes

Equilibrium (solution for the game)	*Strategies chosen by movement*	*Strategies chosen by state*
The game in Figure 3.2	Don't take demonstrative action, or take demonstrative action and back down	Threatens and uses force
The game in Figure 3.3	Take demonstrative action and resist	Threatens and does not use force
The game in Figure 3.5	Take demonstrative action and resist if $p^* > p$ ($p = 50\%$); back down if $p^* < p$ ($p = 50\%$)	Threatens, shows force and uses force if the state is tough; does not use force if the state is humane

Three types of preference structure have been discussed in Figures 3.2, 3.3, and 3.5. What distinguishes the preference structure in Figure 3.2 from the one in Figure 3.3 is the fact that the state has reversed its preferences. In Figure 3.2, the state prefers retention of its power/authority to legitimacy, and in Figure 3.3, it prefers legitimacy to power. What makes the preference structure in Figure 3.5 different from those in Figures 3.2 and 3.3 is that the state holds its preferences as private information. The participants in Figure 3.5 are therefore uncertain about what the state prefers.

These three types of preference structure induce and constrain both the state's and the participants' options during the development of a movement. It is through them that movement outcomes occur. Table 3.4 presents a summary of their specific effects upon the movement's development and outcomes.

Table 3.4 shows the considerable effects of these preference structures on movement development and outcomes. Unfortunately, existing theories in the study of the dynamics of social movements have not paid enough attention to them. Although the internal strategic approach has emphasized the interaction between movements and the state, there are no concepts and techniques in that approach to study the preference structures—the mechanics of interaction.

Of the three types of preference structure, the game with incomplete information is the most important. This chapter has discussed the important features of this game, the sequence of choices, and the alternatives from which each side must choose in that game. All of these factors play an extremely important role in determining movement development and outcomes.

Table 3.5 Comparison between game theoretical approach and other existing approaches

Comparison between:	*Game theoretical approach*	*Other existing theoretical approaches*
Effectiveness of threats made by a state	Very likely to fail to hinder a social movement	Very likely to hinder a social movement
Determinants of the effectiveness of the threats	Preference structures, movement's payoffs, participants' beliefs about the state's preferences	Power contrast • movement: low power • state: high power
Policy implications	If the state wants to avoid any escalation of the conflict • don't make threats or • do make threats but at the same time must reverse the two conditions in the first proposition	If the state wants to avoid any escalation of the conflict • do make threats

Finally, it is necessary to compare the game theoretical approach and the existing approach in studying the effect of a state's threat made after the emergence of a social movement (see Table 3.5).

It is obvious that there is some inconsistency between reality and the explanations provided by existing theoretical approaches. Evidence shows that state-issued threats have often failed to hinder social movements. As an alternative explanation, the game theoretical approach provides a better understanding of the reality and more accurate, and therefore more valuable, information for policy makers. For instance, there are two conditions under which threats fail (see Proposition 1). If the state wants to avoid an escalation of conflict with a movement by making threats, its agents should understand that threats have limitations. There are two alternatives to choose from in avoiding an escalation of conflict. First, do not make any threats; second, reverse the two conditions under which threats would fail.

4 State's sub-optimal strategies

A two-level game

The fact that state-issued threats are usually ineffective in a movement's development is the first property of the newly created social system. This chapter considers the second property of the social system—namely, that the state plays a two-level game after a movement's resistance to a state-issued threat and that its options are only sub-optimal in a two-level game.

After a movement's resistance to a state-issued threat, will the state choose to use force, or not? While the state tries to make its decisions, is there an institutional context that constrains its choices? Are its choices optimal strategies? How do the state's strategies impact subsequent movement development?

On 26 April 1989, for example, the Chinese government sent a signal that the student protests would no longer be tolerated. Use of the word "turmoil" in the official editorial was heavily loaded and was used by the government to justify any crackdown it might choose to make. But the very next day, on 27 April, the government sent out a diametrically opposed signal—namely, that the student protests could be tolerated. Thousands of armed police officers withdrew from their positions in front of the students and more than 200,000 bystanders. Why didn't the government keep its word? What made the authorities back down?

From 16 to 18 May 1989, when more than 1,000,000 people marched in the street for three days, the government re-evaluated its position. In order to control the situation, government officials asserted their strong intention to use force and, on 19 May, sent their strongest signal that they would do so. Declaration of martial law and the deployment of thousands of troops accompanied by tanks had very clearly demonstrated the government's preferences. But, on 20 May, when 1,000,000 people protested against the implementation of martial law, the troops were called off. What were the mechanics of interaction that induced government officials not to use force? Did the government signal that the student protests could be tolerated show the state's preferences? If it did, were the state's preferences unchangeable?

The emergence of a social movement creates a two-level game for the state, and the preference structures in the game induce and constrain the state vis-à-vis its decision-making process, which, in turn, significantly influences the movement's subsequent strategies and the ultimate outcomes of social movements.

A two-level game

As a concept, the two-level game was first used by Putnam as a metaphor for domestic-international interactions:

> The politics of many international negotiations can usefully be conceived as a two-level game. At the national level, domestic groups pursue their interests by pressuring the government to adopt favorable policies, and politicians seek power by constructing coalitions among those groups. At the international level, national governments seek to maximize their own ability to satisfy domestic pressures, while minimizing the adverse consequences of foreign developments. Neither of the two games can be ignored by central decision-makers, so long as their countries remain interdependent, yet sovereign.
>
> (Putnam, 1988:434)

In the two-level game, according to Putnam, national political leaders are simultaneously engaged in two games, against distinct classes of opponents. At the national level, the opponents are domestic groups; at the international level, the opponents are the government's foreign counterparts. The strategy pursued by national political leaders at one level limits the strategies available to them at the other level. Consider each national political leader, then, as a major player that plays two games at the same time.

Putnam (1988:460) wrote that, "Two-level games seem a ubiquitous feature of social life." And Kenneth A. Shepsle and his colleagues (1985) analyzed the strategy of a legislator simultaneously embedded in two games—one in the legislative arena and the other in the electoral arena. Fritz Scharpf (1988) has interpreted macro-economic policy as the joint outcome of two simultaneous games. In one, the government plays against the unions; in the other, it responds to the anticipated reactions of the electorate.

As we see then, the multiple-game nature of states' activity is not special. However, it is worth noting that, unlike other games, a two-level game has two unique features. First, a major player plays two games with two different opponents (or groups of opponents) at the same time. In domestic–international interactions, for example, Robert Strauss (1987:vii) said of the Tokyo Round trade negotiations: "During my tenure as Special Trade Representative, I spent as much time negotiating with domestic constituents (both industry and labor) and members of the U.S. Congress as I did negotiating with our foreign trading partners."

Second, the options available to the major player in one game are limited by the strategies pursued by the major player in the other game. According to Scharpf (1988), for example, when the state fights inflation and unemployment at the same time, a given action for the state is to take a fiscal and monetary policy stance. But the government's interests are vested in using the policy restrictively against inflation, and expansively against unemployment. Therefore, the state's interests are embedded in pursuing opposite strategies in two simultaneous games.

While Putnam has made a significant contribution in using a two-level game to capture an important feature of social life, it is necessary to supplement his theory because of the need to distinguish between two types of two-level games. In Putnam's two-level game, the major player has an intention: to please two distinct classes of opponents. Unfortunately, however, there is a conflict of interests between these two distinct classes of opponents, which creates an ongoing dilemma for the major player. In domestic–international interactions, for example, there is a conflict of interests between domestic constituents and foreign trading partners; in trade relations between the United States and Mexico, for instance, high tariffs versus low ones may represent this conflict of interests between domestic constituents and foreign trading partners. While national political leaders want and intend to please both opponents, the strategy they pursue in one game limits the strategies available to them in the other game. I define Putnam's two-level game as the "intermediary" two-level game.

In the second type of two-level game, the major player has an intention: to please his opponents in one game and suppress them in another. One of the significant differences between the two types of two-level games is that a conflict of interests exists between the two distinct classes of opponents in Putnam's two-level game, but a common interest exists between the two distinct classes of opponents in the second type. Although the major player intends to suppress his opponents in one game in the second type of two-level game, he cannot afford its adverse consequences in the other game—a game in which he intends to please his opponents. As a result, moves that are rational for the major player in one game may be irrational in the other.

Table 4.1 Comparison of the two types of two-level game

Comparison between:	*Intermediary two-level game (Putnam's two-level game)*	*Discriminatory two-level game (my two-level game)*
Relationship between two distinct classes of opponents	A conflict of interests exists between the two distinct classes of opponents	A common interest exists between the two distinct classes of opponents
Relationship between the major player and other players in a two-level game	Major player tries to please both distinct classes of opponents	Major player tries to please his opponents in one game and suppress them in another game

I define the second type of two-level game as the "discriminatory" two-level game because the major player's intention in one game is the opposite of his intention in another game. Table 4.1 compares these two types of two-level games.

The emergence of social movements usually creates both types of two-level games for the state.

> Social movements provide weapons for the powerless. They create milieus for people or groups with few formal institutional ties and for ideas that are not part of the institutional consensus. Social movements tend to question both accepted relations of power and the ideological underpinnings of those relations.
>
> (Rosenthal, Fingrutd, Ethier, Karant, and McDonald, 1985:1022)

This means that there is the intermediary two-level game in which the state plays with distinct opponents (powerless and powerful) that have opposing interests (institutional consensus and non-institutional consensus). Thus, the choice that state officials must make—whether to facilitate or hinder a movement—is limited in one game by their strategies in the linked game, because powerful groups try to protect their own interests.

After the emergence of social movements, however, the state also has to play another two-level game: the discriminatory two-level game. For example, 1989 was a critical period for implementation of Deng Xiaoping's policies of reforms and openness in China. Government leaders, therefore, desperately wanted to attract more foreign investment, technology, management methods, etc. At the international level, government leaders wanted to please the global community in order to promote integration with the worldwide economic environment. At the national level, however, Deng Xiaoping and his colleagues would not risk the authority of the Communist party. There was a common interest between the movement and the international community—namely, a strong demand for human rights improvement and the legitimization of civil organizations, etc. This discriminatory two-level game created a difficult situation for government officials, whose intention to please the international community in one game constrained their strategies in the other game.

In sum, the multiple-game nature of a state's policies is quite common, but a two-level game is not merely the combination of any two games played by the state at the same time. By definition, a two-level game means:

1 that the opponents have opposing interests—such as pro-choice and pro-life movements—and the state tries to please both, or
2 that the state's interests lie in pursuing opposite strategies in two games, such as the Chinese government's intention in 1989 to suppress its opponents at the national level, but please its opponents at the international level.

Only under these two circumstances are the state's options in one game limited by its strategies in another.

The Chinese government played many other games while it was dealing with the student movement in 1989; but, according to the definition, most of them were not part of a two-level game.

In 1989, for example, one of the most significant non-student movement games involved getting control of regional leaders who were ordering the financial institutions in their areas to lend too much money and begin too many economic projects. The government's most important economic task in 1989 was to control the money supply, which meant placing a major constraint on these regional leaders. When government officials then dealt with the 1989 student movement, this game with regional leaders was not a part of a two-level game, because the government had no interest in pursuing opposite strategies when it was engaged in these two games. Repressing the students would signal the presence of a strong central government—a message that could actually contribute to constraining the money supply and inflation.

Another game in 1989 was keeping the military satisfied. This game also was not a part of a two-level game, because the government's interests lay in pursuing the same strategies in both the game with the student movement and the game with the military. Therefore, only the game with the international community became part of the discriminatory two-level game in 1989.

The complexity of choosing a strategy in the two-level game

If a player is engaged in only one game, his options—and the outcomes ensured for him—are a function of the other player's actions. However, for a major player in a two-level game, who is engaged in two games against distinct opponents, his options and outcomes are not only a function of another player's actions, but also a function of his own actions in another linked game. This difference makes things more complicated.

According to Putnam, the unusual complexity of the two-level game is that moves that are rational for a player in one game may be irrational for the same player in another game. In the entanglements of domestic and international politics, Putnam (1988:435) noted: "Any key player at the international table who is dissatisfied with the outcome may upset the game board and, conversely, any leader who fails to satisfy his fellow players at the domestic table risks being evicted from his seat." This "two-game" metaphor captures the dynamics of international negotiations more accurately than any model based on unitary national actors.

Scharpf also noted the complexity of choosing a strategy in a two-level game when he analyzed the economic environment and strategy choices of industrialized countries in the early 1970s:

> In a nutshell, the dominant economic problem of the 1970s was stagflation, the simultaneous occurrence of exceptionally high rates of inflation and levels of mass unemployment previously unheard of in the postwar period. . . . For this combination of cost-push inflation and demand-gap unemployment, governments were able to fight either inflation or unemployment, but not both at the same time. Worse yet in trying to solve one problem they would aggravate the other one.
> (1988:231–2)

This dilemma is always encountered by the state after the emergence of a social movement. In the intermediary two-level game, if the state facilitates the movement, it may suffer the adverse consequences of this strategy (facilitation) in another game. Likewise, in the discriminatory two-level game, if the state hinders movement participants, contradictions may arise between that strategy (hindrance) and other strategies pursued by the state in the other game.

During the 1989 movement, the Chinese government first threatened the students on 26 April. It showed its power to back up this threat on 27 April, and then backed down after the students resisted. On 20 May, the government threatened movement participants for the second time by declaring martial law and making a show of force—deploying thousands of troops with tanks—then backed down again after the participants resisted.

The Chinese government backed down twice because state officials were involved in playing the discriminatory two-level game. Since there was a lot of pressure from the international community, they knew that a military crackdown on the student movement would be followed by economic sanctions from important trade partners. The government could not afford the adverse consequences of military action because China was in the middle of economic reform in 1989, and had an especially urgent need for foreign investment.

As mentioned previously, the Chinese government wanted to please the international community, and the best strategy for this was to avoid any escalation of conflict between the movement and the state, including "Concede." However, the Chinese government could not pursue this strategy in the domestic game because any signal that showed tolerance for the movement would have created a domino effect that might have resulted in a loss of power for the Chinese Communist Party, as happened to its counterparts in East European countries.

In short, it is a complex matter to choose a strategy after a movement has demonstrated resistance to threat, and shows that the state's choices are constrained by either the intermediary two-level game or the discriminatory two-level game, or both. Because of the unusual complexity of the two-level game, the state usually sends out mixed signals and is always cautious not to clearly disclose its preferences.

Incentives for consistency between the two games

Although choosing a strategy in both types of two-level game is unusually complex (i.e., moves that are rational for a player in one game may be irrational for that same player in the other game) "There are powerful incentives for consistency between the two games" (Putnam 1988:434). How can consistency be achieved, given the contrary demands of the two-level game? And from where do these powerful incentives for consistency come?

In both types of two-level game, any given action for a major player is simultaneously a move in two different games, and neither of these games can be ignored. Therefore, the player must balance his interests in the two games by maximizing the sum of his payoffs from both. This generates a powerful incentive for consistency between the two games.

How can consistency be achieved when the strategies between the two games are in conflict? Putnam (1988) has constructed a theory for this in domestic–international interactions by using a concept he calls "win-sets," and by distinguishing involuntary and voluntary defection. His theory works well in the intermediary two-level game.

In an international conference, negotiators representing two countries meet to reach an agreement, subject to the constraint that any agreement must be ratified by their respective countries. Putnam (1988:438) has defined the "win-set" for a given country as the set of all possible international agreements that would "win." By definition, any successful international agreement must fall within the win-sets of each country.

> Thus, agreement is possible only if those win-sets overlap, and the larger each win-set, the more likely they are to overlap. Conversely, the smaller the win-sets are, the greater the risk that the negotiations will break down.
>
> (Putnam, 1988:438)

According to Putnam, involuntary defection reflects the behavior of a negotiator who is unable to deliver on a promise because of failed ratification, whereas voluntary defection means intentionally reneging in the absence of enforceable contracts. The larger the perceived win-set of a negotiator, the more he can be "pushed around" by the other negotiator.

In discussing the determinants of the win-set, Putnam's theory about consistency between the two games helps in understanding how the state makes strategic choices when confronted by the intermediary two-level game, in which the opponents have conflicting interests. However, there is also the discriminatory two-level game, in which the state's interests are embedded in pursuing opposite strategies, and Putnam's theory of consistency unfortunately does not take this circumstance into consideration. Therefore, a supplementary theory of consistency for the discriminatory two-level game is needed, and is constructed here.

A supplementary theory of consistency and sub-optimal strategies

In the discriminatory two-level game, the major player itself has two respective sets of strategies for the two games, instead of one win-set of

Table 4.2 The government's two sets of strategies in the discriminatory two-level game

The government's sets of strategies	*Game with the movement*	*Game with the international community*
The optimal strategy	To threaten first, show force, and then use force if the students resisted the threat	To concede
The sub-optimal strategy	To threaten first, show force, and back down if the students resisted the threat	To threaten first, show force, and back down if the students resisted the threat
The unacceptable strategy	To concede	To threaten first, show force, and then use force if the students resisted the threat

strategies in the intermediary two-level game. During the 1989 Chinese student movement, for example, the government had a set of strategies in its game with the movement, and another set of strategies in its game with the international community (see Table 4.2).

During the 1989 Chinese student movement, the government's optimal strategy in regard to the uprising was to threaten first, show force, and then use force if the students resisted the threat. The government's sub-optimal strategy was to threaten, show force, and back down if the participants resisted the threat. Although the payoffs for the government were very different, both of them were better than the one created by an unacceptable strategy, which was to concede after the emergence of the social movement. At the same time, the government also had optimal, sub-optimal, and unacceptable strategies in its game with the international community.

In the discriminatory two-level game, the strategies in each set or for each game were arranged in order of utility for the state (see Figure 4.1).

Figure 4.1 shows a set of the state's strategies in one game. The utility afforded the state by these strategies diminishes from left to right. Sub-optimal strategy that cannot maximize the state's utility (in the middle) must be distinguished from the optimal strategy that maximizes the state's utility (on the left side). Although the optimal strategy and the sub-optimal strategy in one game bring the state different payoffs and utility, they are definitely better than the unacceptable one in the game. It is in this sense that both the optimal strategy and the sub-optimal strategy in the set are acceptable. Therefore, the state has two acceptable sets of strategies for the two games, respectively (see Table 4.3).

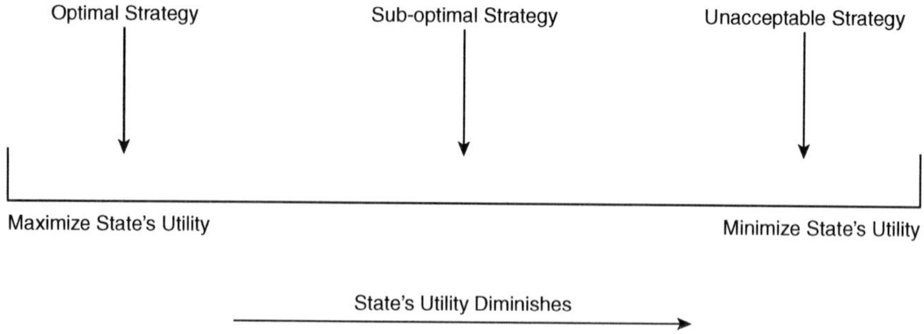

Figure 4.1 A set of strategies in one game

Table 4.3 The government's two acceptable sets of strategies

The government's acceptable sets of strategies	*Game with the movement*	*Game with the international community*
The optimal strategy	To threaten first, show force, and then use force if the students resisted the threat	To concede after the emergence of the movement
The sub-optimal strategy	To threaten first, show force, and back down if the students resisted the threat	To threaten first, show force, and back down if the students resisted the threat

Although a state may have two acceptable sets of strategies respectively for the two games, a simultaneous and positive move for the state in two different games is possible only if those acceptable sets overlap. But the strategies from the overlap cannot be optimal for the state in either game because the state's interests are embedded in pursuing opposing strategies in the discriminatory two-level game (see Figure 4.2).

Figure 4.2 shows how the two acceptable sets of strategies overlap. The strategies on the left side are optimal for the state in one game, and those on the right side are optimal in another game. The left and right sides never overlap. In one game, the state's utility diminishes from left to right; in the other game, it diminishes from right to left. Therefore, the feasible strategies from the overlap are sub-optimal for the state in either game.

During the 1989 Chinese student movement, government officials were confronted with the discriminatory two-level game. The government had to make a move in its game with the movement, and, simultaneously, in its game with the international community. The strategies

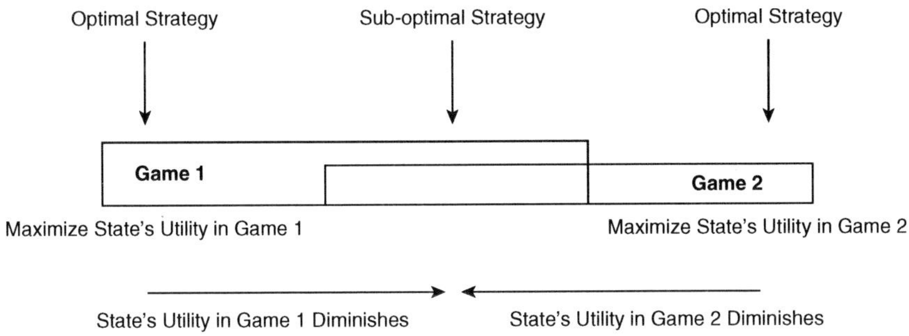

Figure 4.2 Overlapping acceptable sets of strategies

available to the government at that time were not optimal in either game. In fact, the only options included in either game were sub-optimal (i.e., to threaten, to show force, and to back down reluctantly if the students resisted).

In other words, after the movement's resistance to the state's threat, the Chinese government tried not to make a choice between threatening the movement, showing force, and using force (the optimal strategy in its game with movement participants), and conceding, which was the optimal strategy in its game with the international community. One possible choice for the government was a sub-optimal strategy: threatening the movement, showing force, and then backing down if the students and other participants resisted the threat.

In short, in movement development, the discriminatory two-level game as an institutional context affects a state's actions in relation to two aspects. First, state officials want to hold their preferences as private information for as long as possible, because they know that:

1 the international community will be furious if they show that their intention is to suppress the movement by force, and
2 the movement will accelerate wildly if they reveal their intention to please the international community.

Second, although the state chooses a sub-optimal strategy—namely, threatening the movement, showing force, and then backing down—it does so reluctantly and involuntarily because it has run out of other choices.

When a major player chooses a sub-optimal strategy in the discriminatory two-level game, how does his choice affect other players' subsequent strategies? What will happen if state authorities continuously choose sub-optimal strategies as the movement develops?

Significant impacts of the state's sub-optimal strategies on movement development

When state authorities continuously choose sub-optimal strategies, subsequent movement development can be significantly affected in two ways. First, after the emergence of a social movement, movement participants are anxious to know the state's reaction and real preferences before they make any fresh move. If the state continuously plays a sub-optimal strategy, it signals to the movement that the state does not want to use force, that the state is weak, and that government officials are very concerned about the international community's reaction to the military crackdown. And movement participants intend to verify their expectations about the state's preference by using the message sent by the government, and build their beliefs based on the information.

In the Chinese student movement, for example, no one believed that the government would use force after state authorities continuously chose sub-optimal strategies in the discriminatory two-level game. The majority of students and movement participants believed that the state's intention was to please the international community. All evidence has shown that the movement regarded the state's preferences as unchangeable, and all of its signals as reliable.

Unfortunately, while the movement makes all its moves based on the state's intention, there is a possibility that the state may change its preference and play optimal strategies instead of sub-optimal ones. It is therefore very dangerous for a movement to believe that "the changeable government's intention" is unchangeable, and "the temporarily reliable information" is reliable forever. Chapter 5 discusses why and when a state will switch to optimal strategies.

Second, playing sub-optimal strategies deprives the state of the means to actually disclose its real preferences. Although the signals revealed by choosing sub-optimal strategies cannot credibly show the state's preferences in either game, they clearly send a message that the state will back down in certain circumstances. This message generates impacts on the movement's decision-making process, and the state cannot undo these impacts, no matter how hard it tries. As a result, state authorities not only lack the inclination to disclose their preferences in either game—they do not even have the means available to reveal their real preferences when they want to send a clear message to the movement.

Although the state is powerful and is backed by strong military force, this lack of a way to disclose its real preferences may paralyze officials when it comes to controlling the dynamics of social movements. Chapter 6 discusses in detail how the state can lose this control through a lack of means to disclose its preferences.

The following proposition about the state's choice of strategies in the discriminatory two-level game may be drawn from this chapter.

> *Proposition 2: after a movement's resistance to a state-issued threat, the state plays the discriminatory two-level game, in which its options in either game are only sub-optimal: to threaten, to show force, and to back down reluctantly. As a result, the state not only significantly influences the movement's subsequent strategies but also deprives itself of its means to disclose its real preferences.*

Conclusion

The important finding in this chapter is that the emergence of a social movement creates a two-level game for the state: either the intermediary two-level game, or the discriminatory two-level game, and sometimes both. The state has to play two games with two different opponents at the same time. Given the unusual complexity of the two-level game, moves that are rational for the state in one game may be irrational in the other. The reason for this complexity is either that the opponents have distinctly opposite interests, or that the state's interests are embedded in pursuing opposite strategies in the two games. This structure of the two-level game constrains the state's choices and induces state authorities to hold their preferences as private information.

Despite this degree of complexity, powerful incentives exist to establish consistency between the two games because the state has to move simultaneously in two different games, and neither game can be ignored. More specifically, in the discriminatory two-level game, since consistency between the two games is achieved by abandoning the optimal strategy in each game, the only choices remaining to the state are sub-optimal strategies for either game.

The signals disclosed by the state in choosing sub-optimal strategies significantly impact the movement's subsequent strategies because movement participants tend to verify their expectations about the state's preference by using the message sent by the government, and they build their beliefs based on this information. However, it is always possible that the state's preference will change and switch to the optimal strategies. Therefore, the state's preferences, as revealed by its signals, are not unchangeable.

Table 4.4 Comparison between game theoretical approach and other existing approaches

Comparison between:	*Game theoretical approach*	*Other existing theoretical approaches*
What determines whether the state chooses to "facilitate" or "hinder" a social movement?	It is not only the state's own intention, but also the institutional context, such as playing a discriminatory two-level game	It is the state's own intention, since the state is in the position to do whatever it intends to, such as lower or raise the costs of collective action
What determines whether the state is in control of the dynamics of social movements?	It is not only military force, but also the means to disclose the state's preferences to the movement	It is military force
Policy implications	For the movement, it is dangerous to neglect the possibility that the state may change its strategies from sub-optimal to optimal ones	
	For the state, playing sub-optimal strategies may paralyze officials in their efforts to control the dynamics of social movements	

Playing the sub-optimal strategies also creates a situation that deprives the state of a means to disclose its real preferences in the discriminatory two-level game. Although the state's position vis-à-vis movement development is backed up by strong military force, a lack of means to disclose its preference may easily paralyze government officials when it comes to controlling the dynamics of a social movement.

Finally, it is necessary to compare the game theoretical approach and the existing approach in studying the state's strategy choices (see Table 4.4).

As a tool in analyzing the interaction between states and movements, the game theoretical approach is more powerful and complete than the existing approach. The idea that threatening, but not repressing, is a sub-optimal strategy is the key to understanding the state's reaction to a movement. More importantly, by playing the sub-optimal strategy, the state generates a situation that significantly impacts the further development of a social movement.

5 Short-term gain and long-term loss for the participants

The dynamics of repeated games

The newly created social system composed of the movement and the state has a third property—namely, the movement faces a tradeoff between short-term payoffs and long-term interests during its development. What are movement participants' short-term payoffs and the long-term interests? Why is there a tradeoff for the movement between these short-term payoffs and long-term interests?

Chapter 3 has shown that a game with incomplete information symbolizes the institutional context in which the movement chooses its strategies, and Chapter 4 has shown that the existence of a two-level game, as the institutional context, constrains the state's strategies. While Chapters 3 and 4 have focused on the single-shot game—a game played only once—Chapters 5 and 6 will concentrate on repeated games.

Games are considered as repeated when players interact by playing a single-shot game numerous times. Chapter 3, for example, has shown that the movement's development could be modeled as a single-shot game with four steps (see Figure 3.1). In the first step, movement participants decide whether to take demonstrative actions. If they do so, the state has three choices in the second step: to make a concession, to repress the movement, or to threaten the participants. If the state makes a threat, participants have to decide in the third step whether or not to resist. If they do so, the state chooses whether or not to act on its initial threat in the fourth step. If the state chooses not to, the game starts over again (i.e., movement participants will decide whether to take new actions). In the Chinese student movement of 1989, this single-shot game was played three times.

The first round was played in April 1989, after the emergence of the student movement, when Chinese officials sent a signal that the student protests would no longer be tolerated. Officials used the term "turmoil" to justify any crackdown they might choose to make. But when students disregarded this signal and marched in Beijing, the government withdrew thousands of armed police officers from their positions in front of the students and more than 200,000 bystanders.

The second round was played in May 1989, when the movement launched a hunger strike at Tiananmen Square and more than 1,000,000 people marched in Beijing to support it. On 19 May, the government declared martial law and deployed thousands of troops accompanied by tanks. But, when 1,000,000 people protested against martial law, the troops and tanks were again called off.

Finally, in June 1989, after students had debated whether it was time for them to go back to the campuses to resume their studies, another hunger strike—on a much smaller scale than the first one—began at Tiananmen Square. On 3 June, the government announced that "the turmoil had developed into a counterrevolutionary rebellion" (*People's Daily*, 4 June 1989). Loudspeakers near Tiananmen Square, and government-controlled radio and television, repeatedly broadcast an announcement warning people to stay off the street or "bear responsibility for their own fates" (*People's Daily*, 4 June 1989). When students resisted the threat, the massacre began late in the night of 3 June, and the repeated game ended in the morning of 4 June.

In short, there were three rounds of the repeated game in the 1989 Chinese student movement. I define the short-term payoffs as the movement's gains in the first round and the second round. The long-term interests are the movement's payoffs at the end of the repeated game.

It is extremely important to study repeated games for two reasons. First, movement development viewed as repeated games includes many time periods and sets of inter-temporal dependency-of-action. Tirole wrote, "The choices made by the actors in period t affect their set of feasible choices in a future time period t + t´, where t´ > 0" (Tirole 1988: 206). More importantly, the players' payoffs in period t also significantly affect their payoffs in a future time period, t + t´.

Second, repeated games capture the idea that players have to take into account the impact of their current action on the future actions of other players. In the case of the 1989 Chinese student movement, for example, the government's first reaction to the student demonstration was to threaten, to show force, and then to back down. The impact of this strategy was significant: students resisted all threats made by the government in the three-round repeated game. Although Chinese authorities wanted to avoid bloodshed in the third round, government officials could not undo the impact of the strategies they had utilized in the first and second rounds.

Scholars of social movements generally agree that it is very important to study the dynamics of social movements—namely, the changes in the strategies of both the movement and the countermovement. Unfortunately, none of these scholars has ever studied the systematic connection between changes in the movement's payoffs and changes in the state's strategies.

The study of repeated games now provides an opportunity to apply the concept of the mixed strategy Nash equilibrium to the field of social movement. In game theory, there are two kinds of Nash equilibriums. One is pure strategy Nash equilibrium, in which the equilibrium strategies are played with probability 100 percent. The other is mixed strategy Nash equilibrium, in which a probability distributes over all the strategies, such as playing one strategy with probability 25 percent and playing another with probability 75 percent.

By utilizing the concept of the mixed strategy Nash equilibrium, systematic analysis of how state strategies respond to changes in the movement's payoffs will significantly deepen our understanding of the dynamics of social movements. More importantly, it will unfold why and how the movement's short-term payoffs affect its long-term goal.

Although there are many important issues in the study of the repeated games, this chapter is only concerned with two issues that are related, respectively, to the first proposition in Chapter 3 and the second proposition in Chapter 4.

The first proposition (in Chapter 3) specifies two conditions under which threats made by the state are likely to fail in their objectives. The first of these is when participants believe that their chances are 50/50, or less than 50 percent, that the state will make good on its threats. The second condition occurs when the movement's payoff from resisting the threat and not being repressed is greater than that from backing down. In this chapter, one of the main questions asked is whether these conditions change in the repeated game. If they do, why and how have they changed? More specifically, will participants' beliefs about the probability that the state will make good on its threats change? Will participants' payoffs increase or decrease in the repeated games?

The second proposition (in Chapter 4) specifies that, in a two-level game, the state's options in either game are only sub-optimal strategies: to threaten, to show force, and to back down reluctantly. In this chapter, then, one of the challenges is to find out whether the state's options in a two-level game have changed in the repeated game. If so, what changes them, and in what ways do they change?

The participants' changing beliefs: diminishing dangers

In the first round of the repeated game, the state holds its preferences as private information, and participants believe that the chances are 50/50 that the state will make good on its threats. Does this belief change during the movement's development? How do participants form or modify their beliefs in the repeated game?

This section answers these questions by introducing a learning process based on Bayes' rule, which is a consequence of the axioms of probability theory. This process describes how participants draw inferences and form beliefs about the state's preferences by observing the state's words and deeds, and how participants modify their beliefs during the movement's development.

Bayes' rule defines the relationships among three variables. The first variable is the participants' prior attitude toward the state's preferences: what they believe before the state takes any action. After the state has threatened the movement, the participants—who are ignorant of the state's preferences—are confronted with the hypothesis that the state is humane. By definition, this is true if the state prefers legitimacy to power (i.e., the state does not use force after participants resist its threat). Is this hypothesis true or false? If it is false, then the state is, by definition, tough, because it prefers retention of its power/authority to legitimacy (i.e., the state uses force after participants resist its threat). Each movement participant has beliefs about the truth or falsehood of this hypothesis. Most believe in a 50/50 chance that the state is humane. This, then, is the participants' prior belief.

The second variable is new information that arises. As mentioned earlier, when the movement develops, the four-step single-shot game is repeated. Each time this happens, movement participants receive some data by observing the state's words and deeds—data that may be relevant in assessing the truth of the hypothesis that the state is humane. This provides new information for the participants.

The third variable is the participants' posterior beliefs about the state's preferences. After they receive the new information, they revise their initial beliefs and form their posterior beliefs. Suppose that the single-shot game is repeated once in every round. The participants' posterior beliefs in the first round become their prior beliefs in the second round, and so on. In each round, prior beliefs and new information are the two independent variables that determine posterior beliefs.

How do posterior beliefs depend on prior beliefs and new information? What can we expect when participants revise their prior beliefs about the state's preferences in light of the new information? Bayes formalizes the answers to these questions, according to Bayes' rule:

$$P(H|D) = \frac{P(D|H)P(H)}{P(D|H)P(H) + P(D|-H)P(-H)} \quad (5)$$

where P(H) and P(–H) denote the first variable: the participants' prior beliefs. P(H) represents prior beliefs about the probability that the hypothesis, H, (the state is humane) is true, and P(–H) is the probability that the hypothesis is false—namely, that the state is tough; P(H) + P(–H) = 1.

P(D|H) and P(D|–H) denote the second variable: the new information. P(D|H) is the probability that one kind of new data will be observed if the hypothesis, H, is true; and P(D|–H) is the probability that the same data will be observed if the hypothesis, H, is false.

P(H|D) denotes the third variable: the participants' posterior beliefs. P(H|D) is the probability that hypothesis H is true, conditional upon observation of the new data, D.

More specifically, in the model presented by Figure 3.5, in Chapter 3:

P(H) = probability that the state is humane
P(–H) = probability that the state is tough
P(D|H) = probability that participants would observe the data: the state will not use force if it is humane
P(D|–H) = probability that participants would observe the data: the state will not use force if it is tough
P(H|D) = probability that the state is humane, conditional upon having observed that the state does not use force.

Does P(D|–H) ever exist? In other words, if the state is tough, it should use force; but why will participants observe the fact that the state does not use force? According to the second proposition in Chapter 4, the options available to both the humane state and the tough state in the discriminatory two-level game are only sub-optimal. This means that, as long as the state plays a two-level game, it is very difficult for it to make a choice between:

1 threatening the movement, showing force, and using force if the participants resist; and
2 not threatening the movement at all.

This is true because both choices are optimal strategies in the discriminatory two-level game. For both the humane state and the tough state, the only choice is to threaten, to show force, and not to use force if the participants resist. Therefore, P(D|–H) does exist.

Although Bayes' rule provides a formula—see equation (5)—to show how participants' posterior beliefs depend on their prior beliefs and the new information they receive, it is not easy to understand the relationship between the dependent variable and the independent variables. To do so, it is necessary to make a change in equation (5):

$$\because\ P(-H|D) = 1 - P(H|D)$$

$$\because\ P(H|D) = \frac{P(D|H)P(H)}{P(D|H)P(H) + P(D|-H)P(-H)}$$

$$\therefore\ P(-H|D) = \frac{P(D|-H)P(-H)}{P(D|H)P(H) + P(D|-H)P(-H)} \tag{6}$$

$$\therefore\ \frac{P(H|D)}{P(-H|D)} = \frac{P(H)}{P(-H)} \times \frac{P(D|H)}{P(D|-H)} \tag{7}$$

On the left side of equation (7), $\frac{P(H|D)}{P(-H|D)}$ are the posterior odds that the hypothesis, H, is true, conditional upon having observed the data, D. As the posterior odds increase, the probability increases that H is true. In other words, the greater the posterior odds, the more likely the participants are to believe that the state is humane. The smaller the posterior odds, the more likely the participants are to believe that the state is tough.

On the right side of equation (7), $\frac{P(H)}{P(-H)}$ are the prior odds that the hypothesis, H, is true. $\frac{P(D|H)}{P(D|-H)}$ represent the likelihood ratio that the data, D, would be observed if H were true rather than false. The greater the likelihood ratio, the more likely it is that the hypothesis, H, is true.

The relationship between the dependent variable (the posterior odds), and the independent variables (the prior odds and the likelihood ratio) is much easier to understand in equation (7) than it is in equation (5). In equation (7), the participants' posterior beliefs can be found by simply multiplying participants' prior beliefs by the likelihood ratio.

What would we expect to see when participants revise their prior beliefs in light of the new information? There are three directions that

Table 5.1 Three possibilities to revise prior beliefs

First independent variable: prior odds	*Second independent variable: likelihood ratio*	*Dependent variable: posterior odds*
Given P(H)/P(–H)	If P(D|H) > P(D|–H)	Then posterior odds > prior odds
Given P(H)/P(–H)	If P(D|H) = P(D|–H)	Then posterior odds = prior odds
Given P(H)/P(–H)	If P(D|H) < P(D|–H)	Then posterior odds < prior odds

participants' posterior beliefs can take in response to new information (see Table 5.1).

In Table 5.1, if the data are more likely to appear when the hypothesis is true than when it is false—i.e., P(D|H) > P(D|–H)—then the participants' posterior odds are greater than their prior odds. If the data are just as likely to appear whether the hypothesis is true or false—i.e., P(D|H) = P(D|–H)—then the participants' posterior odds are equal to their prior odds. If the data are more likely to have occurred because of a false hypothesis—i.e., P(D|H) < P(D|–H)—then the participants' posterior odds are smaller than their prior odds.

Although both the participants' prior beliefs and the new information they receive determine their posterior beliefs, as the Bayesian learning process demonstrates, the possibility that the posterior odds will increase or decrease compared to the prior odds depends solely on the likelihood ratio: the new information which the participants received.

What is the new information that participants receive in the first and second rounds of the repeated game? According to the second proposition (in Chapter 4), the only choice the state has is to threaten, to show force, and not to use force if the participants resist. Therefore, the new information that the participants receive is that the state did not use force.

How does one assess P(D|H) and P(D|–H) based on the new information? First, participants recognize that the data, D, clearly show that the state did not use force, regardless of the hypothesis. The second step is that participants combine the data, D, with the hypothesis, based on common knowledge. It is common knowledge, for example, that humane states do not use force, although some tough states may choose not to use force in certain circumstances, such as in the discriminatory two-level game. In other words, for the participants, P(D|H) is certainly greater than P(D|–H), in both the first and second rounds. Therefore, the likelihood ratio in both rounds is greater than 1.

Table 5.2 shows how posterior odds change, based upon changes in the prior odds and the likelihood ratio.

Table 5.2 Increase in the posterior odds

Repeated game	*Prior odds*	*Likelihood ratio*	*Posterior odds*
The first round	Prior odds = 50/50 = 1	Likelihood ratio > 1	Posterior odds > prior odds
The second round	Prior odds >1	Likelihood ratio > 1	Posterior odds > prior odds

As mentioned above, the greater the posterior odds, the more likely the participants are to believe that the state is humane. In the first round of the repeated game, participants believe in a 50/50 chance that the hypothesis is true—namely, that the state is humane. In the second round, they believe that the chance is greater than 50 percent that the hypothesis is true. While participants' belief that the hypothesis is true gradually increases, their belief that the state will use force decreases. As a result, participants believe that the danger to them is decreasing during movement development. This is the first dynamic characteristic that emerged from the repeated games.

The participants' increasing payoff, X3

It has been stated above that there are two conditions under which threats made by the state are likely to fail in their objectives. The first of these is when participants believe that the chances are 50/50, or less than 50 percent, that the state will make good on its threats. The second condition occurs when the movement's payoff from resisting the threat and not being repressed is greater than that from backing down.

The previous section discussed how the first condition changes in the repeated games: participants believe that the chances that the state will make good on its threats are not only less that 50 percent, but are also decreasing. Now I will examine the second condition: in repeated games, is the movement's payoff from resisting the threat and not being repressed always greater than that from backing down? In order to answer this question, I will first review the movement's payoff from resisting and not being repressed, and then, in the next section, discuss the movement's payoff from its backing down.

The participants' payoff from resisting and not being repressed is X3. In Table 3.3, it was shown that X3 = G + H. Here, G denotes a substantial gain for the participants (such as human rights improvement or legitimization of civil organizations, etc.), G > 0; and H denotes honor to the participants who resisted the threat, and G > H. When participants resist and the state backs down, they reap both substantial gains and honor: G + H.

There are two ways to find out how X3 changes as the movement develops. The first is to check X3 at the average—i.e., X3 is the per capita payoff from resisting and not being repressed. The second way is to check X3 as a whole—i.e., X3 is the gross payoff for participants who resist and are not repressed.

Here, there is no difference between the two ways. This holds true because one of the important assumptions in the study, presented in Chapter 1, is that all of the participants are wholly unified as a single individual. Therefore, no quantitative difference exists here between the per capita payoff and the gross payoff.

What are the circumstances under which X3 changes? It is the repeated games in which a single-shot game with four steps is repeated: movement participants take action, authorities threaten the movement, participants resist, the state backs down, and so on. Suppose, then, that the single-shot game is repeated n times (n = 1, 2, 3 ...). $X3_1$ ($X3_1 = G_1 + H_1$) is the participants' payoff from resisting and not being repressed in the first round, $X3_2$ ($X3_2 = G_2 + H_2$) is the payoff in the second round, and $X3_n$ ($X3_n = G_n + H_n$) is the payoff in the nth round. In order to find out how X3 changes in the repeated games, it is necessary to compare G_n with G_{n+1} and H_n with H_{n+1}, respectively.

Is G_{n+1} always greater than G_n? In other words, are the participants' substantial gains in the (n+1)th round always greater than those in the nth round? Yes, they are, and two pieces of factual evidence support this answer. The first concerns the substantial gains of the movement's original participants, and the second concerns the gains to new participants.

First, the original participants' substantial gains escalate in the repeated games. Social movements occur because participants ask to change the allocations of power and rights in a social system. By nature, then, movement development is a process in which power and rights have been reallocated between the state and the movement participants. The escalation of gains to the original participants is a direct result of this reallocation.

When the state backs down for the first time after participants resist its threats, it loses power. The more power the state authorities lose, the greater the political gain to movement participants. After the state backed down in the first round of the 1989 Chinese student movement, the government was forced to negotiate with an unofficial student organization for the first time since the Chinese Communist Party had gained control of China in 1949. In effect, students won the right to build their own organizations, which had been illegal before.

In the second round, a state again attempts to induce participants to back down by making a harsher threat, which shows that state officials want to regain their power and force movement participants to give up the political rights they gained in the first round. Of course, the participants do not want to give up these rights—their only concern is to demand additional political rights that can effectively protect the rights they won in the first round. If the state does not use force in the second round, it loses even more power, and movement participants succeed in securing the political rights they asked for to protect the rights they won in the first round. This shows how the original participants' substantial gains escalate in the repeated games.

In 1989, when the state increased the degree of seriousness of its threat in the second round by declaring martial law, students and other movement participants began to question the government's fundamental legitimacy. Then the state backed down again and, after it caved in, the original participants' substantial gains escalated because they came to hold the right to revoke government authority. They openly criticized some of the top leaders and asked them to resign.

Second, new participants added their gains into G. When the state yielded after participants resisted its threats, even more people joined the movement. Some of them simply showed their support for the movement by participating in it, while others participated to secure their own particular political demands.

In April 1989, 95 percent of the participants in China's democratic movement were students; in May 1989, intellectuals and factory workers joined the movement. Then other city dwellers in Beijing also took part. After the government backed down for the second time, G_2 denotes the increasing gains that the government had to reckon with, including with an unofficial student organization and unofficial associations representing intellectuals, workers, and other citizens. All of these associations were new, and had been illegal before.

In short, G_{n+1} is always greater than G_n in the repeated games, due to the two reasons discussed above.

Is H_{n+1} always greater than H_n? In other words, do public bystanders always show more honor to participants in the $(n + 1)^{th}$ round than they do in the nth round? The answer is no. In the repeated games, there is no significant difference between H_n and H_{n+1} (i.e., $H_n \approx H_{n+1}$). When participants resist after the state threatens, they run a risk because the state may repress the movement by force. The greater the risk, the greater the honor bestowed upon participants by bystanders, who appreciate their bravery and courage.

Risk is determined by two factors. The first is the cost to participants if the state resorts to violence: the greater the cost, the greater the risk. The second factor is the probability that the state will instigate violent repression of the movement. As discussed earlier, this possibility is private information held by the state, but participants and bystanders have their own beliefs about it. If they believe that chances that the state will use force have increased, the perceived risk becomes greater.

In the previous section it was shown that, in repeated games, participants' and bystanders' belief in the likelihood that the state will resort to violent oppression gradually decreases. Therefore, the risk run by participants should decrease. In practice, however, the risk does not decrease because the cost to those participants gradually increases. In repeated games, threats made by the state escalate round by round. The state attempts to induce participants to back down by threatening them. If the threat made in the first round does not work, the state must make the second threat harsher than the first. The harsher the threat, the greater the implied cost to participants.

There is no significant difference, then, between H_n and H_{n+1}, because H is determined by the risk to participants, and risk is a function of the cost to participants and the probability that the state will use force. In the repeated games, the cost increases, but participants decreasingly believe that the state will use force. To some degree, therefore, changes in these two variables cancel each other out in terms of impacting H. This discussion is summarized in Table 5.3, below.

After comparing G_n with G_{n+1} and H_n with H_{n+1}, respectively, we conclude that the participants' payoff, X3, from resisting and not being repressed, increases in the repeated games, as follows:

$$\because X3 = G + H,\ G_{n+1} > G_n, \text{ and } H_{n+1} \approx H_n$$
$$\therefore X3_{n+1} > X3_n \qquad (8)$$

The participants' payoff, X3, from resisting and not being repressed, is gradually increasing, and that becomes the second dynamic characteristic of repeated games.

Table 5.3 Constant value of H (honor) in the repeated game

Repeated game	*Cost*	*Probability according to beliefs*	*H (honor)*
n^{th} round	⇗	⇘	$H_n \approx H_{n-1}$
$(n+1)^{th}$ round	⇗	⇘	$H_{n+1} \approx H_n$
$(n+2)^{th}$ round	⇗	⇘	$H_{n+2} \approx H_{n+1}$

The participants' constant payoff, X2

The participants' payoff from backing down after the state shows force is X2. It has been shown in Table 3.3 that X2 = –L. –L denotes a general loss to participants because they lose face when they back down in the face of state-issued threats and $-L < 0$. After participants back down, bystanders blame them and participants lose their reputation. In order to find out how X2 changes in the repeated games, it is necessary to compare $-L_n$ with $-L_{n+1}$. In other words, do bystanders blame participants to a greater or lesser degree if they back down in the (n+1)th round instead of in the nth round?

–L is determined by two factors that vary in the repeated games. The first factor is how serious the state-issued threats are. The harsher these threats, the less participants are blamed by bystanders after the movement backs down, because no one wants participants to put their lives in danger. Since the state wants to make its threats effective, state-issued threats in the (n+1)th round (after participants have resisted in the nth round) are always harsher than those in the nth round. As a result, the participants' payoff X2 from backing down increases: $-L_n < -L_{n+1}$.

The second factor, as the public sees it, is the probability that the state will use force. If this probability declines, bystanders place more blame on participants after the movement backs down. As discussed in the first section of this chapter, participants' belief that the state will use force gradually decreases in the repeated game. As a result, the more blame on participants, the fewer payoffs they have. The participants' payoff, X2, from backing down declines: $-L_n > -L_{n+1}$.

To summarize, there is no significant difference between $-L_n$ and $-L_{n+1}$ because –L is determined by the degree of seriousness of the state-issued threats and the probability that the state will use force. Although these two factors vary in repeated games, the impact on –L of the changes in the two factors cancel each other out to some extent. This discussion is summarized in Table 5.4.

Table 5.4 Constant value of the participants' payoff, X2

Repeated game	*Degree of seriousness of the state-issued threats*	*Probability according to beliefs*	*–L (general loss)*
nth round	⇗	⇘	$-L_n \approx -L_{n-1}$
(n+1)th round	⇗	⇘	$-L_{n+1} \approx -L_n$
(n+2)th round	⇗	⇘	$-L_{n+2} \approx -L_{n+1}$

Table 5.5 All changes in the participants' beliefs and payoffs in the repeated game

Repeated game	*The participants' beliefs that the state will use force*	*The participants' payoff, X3*	*The participants' payoff, X2*
n^{th} round	↘	↗	Constant
$(n+1)^{th}$ round	↘	↗	Constant
$(n+2)^{th}$ round	↘	↗	Constant

The third dynamic characteristic of the repeated game is that the participants' payoff, X2, from backing down after the state shows force, remains constant to some degree.

Decreasing the probability of the state choosing the sub-optimal strategy

The first proposition in Chapter 3 specifies two conditions under which participants always resist threats in the single-shot game. In the repeated game, state-issued threats are also very likely to fail because, as the three sections above demonstrate, these two conditions are also satisfied in the repeated game:

1 participants believe that the chances are 50/50, or less than 50 percent, that the state will make good on its threats; and
2 the movement's payoff from resisting and not being repressed, X3, is greater than that from backing down, X2.

Table 5.5 summarizes all of the changes in participants' beliefs and payoffs in the repeated game.

So far, however, all discussion in this chapter has been about movement participants. What about the state?

The second proposition (in Chapter 4) specifies that, in a discriminatory two-level game, the state's options are always sub-optimal: to threaten, show force, and back down reluctantly. But the game described in the second proposition is the single-shot two-level game. When a movement develops, will the state switch to optimal strategies? If it does, when, how, and why will it happen?

In the case of the 1989 Chinese student movement, the Chinese government decided to choose one of the optimal strategies in the last round of the repeated two-level game—namely, "to threaten and to use force"—instead of the sub-optimal strategy, "to threaten and back

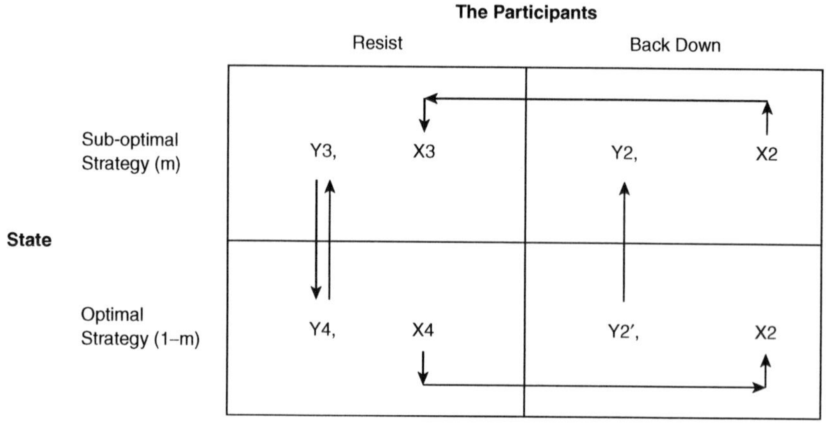

Figure 5.1 A static game without a pure strategy Nash equilibrium

down." What changed the state's options? And, more importantly, do all the changes in participants' payoffs impact the state's strategies?

When the state changes its strategies from sub-optimal to optimal strategies, it demonstrates that the repeated game has the fourth dynamic characteristic—namely, a decreasing probability that the state will choose sub-optimal strategies. What makes this probability decrease? Is there any connection between these changes in the probability, and changes in the participants' payoffs?

Game theory provides a powerful tool—the mixed strategy Nash equilibrium—for study of the connection between changes in the state's strategies and changes in the participants' payoffs. For any game in which each player would like to outguess the other, there is no pure strategy Nash equilibrium. This is due to the fact that, given any combination of pure strategies, one player is always better off changing to the other strategy. The only Nash equilibrium in the game, therefore, is the mixed strategy equilibrium.

In 1989, after the Chinese government threatened the movement and showed force, both the participants and the state had two choices. Participants could resist the threat or back down, while the state played either the sub-optimal strategy (backing down), or the optimal strategy (using force). In this game, the participants would have liked to know whether the state would use force, and state officials wanted to know whether movement participants would resist threats. Figure 5.1 shows this situation in a static game to which the mixed strategy Nash equilibrium can be applied.

The mixed strategy Nash equilibrium in this game has two advantages for the study of the fourth dynamic characteristic of the repeated game. First, it builds connections between the probability that the state chooses sub-optimal strategies and participants' payoffs, and it provides an opportunity to study how changes in participants' payoffs make the probability decline. Second, the mixed strategy Nash equilibrium generates new information that can be used to describe various ways in which the state responds to changes in the movement's payoffs.

Y2, X2, Y3, X3, Y4 and X4 denote the same payoffs as in Figure 3.3 in Chapter 3. Y2′ is the state's payoff from using force if officials succeed in getting participants to back down. The chance that the state will play sub-optimal or optimal strategies is private information held by the state, but participants have their own beliefs about it. Participants believe that the state will play the sub-optimal strategy (to threaten and back down) with probability m, and play the optimal strategy (to threaten and use force) with probability 1-m.

In Figure 5.1, the participants' ordinal utility is common knowledge—i.e., U (X3) > U (X2) > U (X4). Although the state's preferences are still private information, it is known that the tough state's ordinal utility is U (Y2) > U (Y2′) > U (Y4) > U (Y3). The humane state's ordinal utility is U (Y2) > U (Y2′) > U (Y3) > U (Y4). (All explanations can be found in Chapter 3.)

In the case of the 1989 Chinese student movement, was the Chinese government tough or humane? The movement participants were uncertain. Therefore, in the game in Figure 5.1, there is no pure strategy Nash equilibrium, and there is a mixed strategy Nash equilibrium. (The proof is given in the Appendix.)

$$m = \frac{X2 - X4}{X3 - X4} \tag{9}$$

$$1 - m = \frac{X3 - X2}{X3 - X4} \tag{10}$$

The mixed strategy Nash equilibrium clearly shows there is a connection between the probability, m, and participants' payoffs in the repeated two-level game. In equation (9), there are three variables (X2, X3, and X4) that determine the probability, m. All of them are the participants' payoffs. Taking a comparative static approach, we can examine the

Table 5.6 Comparative static based on the static game in Figure 5.1

Independent variables		*Dependent variable*	
X2	⇘	m	⇘
X3	⇗	m	⇘
X4	⇗	m	⇘

effect of participants' payoffs (X2, X3, and X4) on the probability, m, that the state will play the sub-optimal strategy:

$$\because m = \frac{X2 - X4}{X3 - X4}$$

$$\therefore \frac{\partial m}{\partial X2} > 0, \frac{\partial m}{\partial X3} < 0, \frac{\partial m}{\partial X4} < 0 \qquad (11)$$

The comparative static results are summarized in Table 5.6.

As mentioned above, the second dynamic characteristic of the repeated game is that the participants' payoff, X3, increases. In Table 5.6, we see that increasing X3 makes the probability, m, decrease. The third dynamic characteristic of the repeated game is that the participants' payoff, X2, remains constant. In Table 5.6, if X2 declines, the probability, m, also declines. But the constant value of X2 does not impact the probability, m.

The participants' payoff from being repressed after they resist is X4. It has been shown in Table 3.3 that X4 = –D + H. Huge loss, including death for the participants, is denoted by –D, and D > 1. Honor to the participants who have resisted the threat before the state uses force is denoted by H. When participants resist and the state backs down, participants reap substantial gain, as well as honor. In the second section of this chapter, it has been demonstrated that H is constant in the repeated game. Therefore, only –D determines whether X4 increases in the repeated game.

In the repeated game, –D is decreasing—i.e., the participants' payoff, X4, is declining as a result of repression by the state, because the state's threats escalate round by round. The state attempts to induce the participants to back down by threatening them. If the threat in the first round does not work, the state must make the second threat harsher. The harsher the threat, the bigger the loss is for the participants. The bigger the loss, the smaller is the participants' payoff (X4) from being repressed. In Table 5.6, if X4 increases, then the probability, m, declines. In the repeated game, however, X4 declines and the probability, m, increases.

Table 5.7 Increase in the probability that state threatens and uses force

Repeated game	*Movement's payoff from resisting and not being repressed: X3*	*Movement's payoff from being repressed after resisting: X4*	*Probability that state threatens and backs down: m*	*Probability that state threatens and uses force: 1 – m*
n^{th} round	↗	↘	↘	↗
$(n+1)^{th}$ round	↗	↘	↘	↗
$(n+2)^{th}$ round	↗	↘	↘	↗

How does the probability, m, vary if the participants' increasing payoff, X3, makes the probability, m, decline, and the participants' decreasing payoff, X4, makes the probability, m, increase? The best way to answer this question is to compare the rate at which changes in X3 affect the probability, m, with the rate at which changes in X4 affect the probability, m:

$$\because m = \frac{X2 - X4}{X3 - X4}$$

$$\frac{\partial^2 m}{\partial X3^2} > 0, \quad \frac{\partial^2 m}{\partial X4^2} < 0 \qquad (12)$$

In equation (12), the increase in X3 makes the probability, m, decrease at the increase rate, and the decline in X4 make the probability, m, increase at the decrease rate, which means that X3 has a greater impact on the probability, m, than does X4 (see Table 5.7).

Table 5.7 shows that changes in the participants' payoffs obviously impact the probability that the state will play the optimal strategy. In other words, the fourth dynamic characteristic of the repeated game—the probability that the state will play the sub-optimal strategy decreases—is a direct result of the other dynamic characteristics of the repeated game.

Short-term gain and long-term loss for the participants

In the preceding section, the probability that the state will play the sub-optimal strategy is determined, not only by the state's payoffs, but also by the participants' payoffs. According to game theory, the state's mixed strategies in Figure 5.1 are calculated to neutralize the participants' choice of strategies, not to maximize the state's payoffs. How could the state change the probability of playing sub-optimal strategies, when doing so is not to its advantage?

A significant difference between a mixed strategy Nash equilibrium and a pure strategy Nash equilibrium is that a player's strategy is determined by another player's payoffs in the mixed strategy Nash equilibrium, and by the player's own payoffs in the pure strategy Nash equilibrium. "For this reason, many people find the idea of a mixed strategy Nash equilibrium incredible" (Kreps, 1990a:408).

While the state's strategy keeps movement participants indifferent to their own strategies (to resist or to back down), any changes in the participants' payoffs definitely concern the state. In the repeated game, therefore, the probability that the state will choose the sub-optimal strategy responds to changes in the participants' payoffs.

Although the probability that the state will play the sub-optimal strategy responds to changes in all participant payoffs (X3 and X4), two types of response are observed: elastic and inelastic. The state's response is "elastic" if the probability that the state will play the sub-optimal strategy responds substantially to changes in the participants' payoffs. For instance, when the participants' payoff, X3, changes, the probability changes at the increase rate. On the other hand, the state's response is "inelastic" if the probability that the state will play the sub-optimal strategy responds only slightly to changes in the participants' payoffs. For instance, when the participants' payoff, X4, changes, the probability changes at the decrease rate.

In the newly created social system, which is composed of movement participants and the state, special attention should be paid to the state's elastic response, because the probability that the state will use force significantly increases when the participants' payoffs have increased. The state's elastic response is therefore a key factor in determining the outcomes of social movements. In the 1989 Chinese student movement, for example, while the movement's payoffs (substantial gains and honors) increased, the state's elastic response was to change its strategy in the two-level game from the sub-optimal to the optimal strategy—namely, to suppress the movement by using force.

In short, an important finding has emerged from the study of the systematic connection between changes in the movement's payoffs and changes in the state's strategies: the more gains that movement participants reap in the repeated game, the more likely the state is to use force, regardless of the fact that the participants' belief that the state will use force has declined.

This finding is important because movement participants barely notice the systematic connection between the increase in their payoffs and an escalation in the probability that the state will use force. While participants enjoy the increasing gains and honors they have received, they fail to recognize the increasing possibility that the state will use force. In the case of the 1989 Chinese student movement, for example, the military crackdown occurred while students were confidently expecting the Chinese government to back down, as it had before. Most of the students who were killed believed that the People's Army would never hurt them.

More specifically, in 1989 China, movement participants in the repeated game enjoyed and celebrated short-term gains, such as honor and the legitimization of civil organizations, and then suffered long-term loss when the movement was suppressed by military force. The dynamics of the repeated game provide a tradeoff between short-term payoffs and long-term interests, and this tradeoff becomes the third property of the newly created social system.

This property of the newly created social system provides an alternative for movement participants: if they want to achieve the long-term goal, they must give up some short-term gains. In other words, if the participants' payoff, X3, from resisting and not being repressed remains constant, then the chance that the state will use force does not increase.

During the 1989 Chinese student movement, participants frequently debated whether to give up some short-term gains in order to achieve the long-term goal of working toward the country's political modernization in democratic and legal ways. Unfortunately, all debates ended with an obsession with gaining short-term payoffs. On 27 May 1989, for example, at a meeting organized by the Joint Federation of All Circles in Beijing, all leaders voted that the students should leave Tiananmen Square. They planned to hold a large-scale demonstration to celebrate the victory of the student movement for democracy, and then planned to withdraw from the square on 30 May. If the movement had followed this plan, the government would not have changed its strategy from the suboptimal to the optimal one and the Tiananmen tragedy would never have occurred. However, student leaders from the General Headquarters of Tiananmen Square soon changed their minds. They were supported mainly by non-Beijing college students who had arrived in Beijing only a few days earlier and were eager to see more action. This evidence demonstrates that many of the movement participants, including some of the student leaders, failed to accept the reality that they faced a tradeoff between short-term payoffs and long-term interests during the movement's development. The more they gained in the short term, the more they lost in the long term.

The following proposition about strategies chosen by the state and the participants in the repeated game is drawn from this chapter:

> *Proposition 3: in a repeated game, movement participants decreasingly believe that the state will use force, and their payoff from resisting and not being repressed increases, while their payoff from backing down remains constant. As a result, the movement's anti-threat resistance continues in the repeated game. At the same time, the probability that the state will play the sub-optimal strategy decreases. Therefore, the movement faces a tradeoff between short-term payoffs and long-term interests.*

Conclusion

A central issue in this chapter is to show how a movement outcome, especially an unintended outcome, is embedded within the movement's development. In the repeated game, there are four dynamic characteristics: the appearance of diminishing danger to participants; participants' increasing payoff, X3; participants' constant payoff, X2; and a decrease in the probability that the state will choose the sub-optimal strategy. Together, these dynamic characteristics generate a tradeoff between short-term payoffs and long-term goals for the movement, which is the third property of the newly created social system. If movement participants fail to recognize the tradeoff and obsess over short-term payoffs, unintended outcomes of social movements occur.

The mixed strategy Nash equilibrium generates valuable information about various ways in which the state responds to changes in the participants' payoffs. Among them, the elastic form of response is extremely important because it embeds the unintended outcomes of movements in the development of social movements.

In this chapter, I have applied four principles—Bayes' rule, mixed strategy Nash equilibrium, a comparative static approach, and elasticity as a general term for a rate of change—to the study of movement development. All efforts are goal-directed: "first, the development of a formal apparatus to make the transition between interdependent actions and their systemic consequences, and second, the demonstration that this can lead to a balance or equilibrium at the system level" (Coleman, 1994: 32). The mixed strategy Nash equilibrium in the repeated two-level game is a final coherent state of balance. The new concept—the tradeoff between short-term payoffs and long-term interests for the movement—represents the transition from the movement's and the state's interdependent actions to their systemic consequences.

Finally, it is necessary to compare the game theoretical approach and the existing approach to the study of movement development (see Table 5.8).

This comparison shows that the game theoretical approach has significant advantages for the study of dynamics of movement development. First, the language of game theory (such as "repeated games")

Table 5.8 Comparison between game theoretical approach and other existing approaches

Comparison between:	*Game theoretical approach*	*Other existing theoretical approaches*
What determines participants' beliefs about the state's strategies?	Participants form their beliefs according to Bayes' rule	Participants form their beliefs by observing the state's words and deeds
What determines the state's strategies?	Not only the state's payoffs but also the participants' payoffs	The state's interests; never study the systemic connections between the state's strategies and the participants' payoffs
How does the state respond to the changes in the participants' payoffs?	The state's elastic response to the changes in X3 The state's inelastic response to the changes in X4	Never distinguish the state's elastic response from its inelastic response
Policy implications	For the movement: • it is important to notice the state's elastic response to any changes in X3 • it is extremely important to understand that they face a tradeoff between short-term payoffs and long-term goals For the state: • it is important to realize that its strategies in the first round have a great impact on the future action of participants	

and the Bayesian learning process make it possible for scholars of social movements to study all of the dynamic characteristics of movement development. Second, the modeling techniques in game theory offer powerful tools (such as mixed strategy equilibrium) for systematic study of the connection between changes in one player's payoffs and changes in another player's strategy. As a result, the impact of one player's action on the future actions of another player becomes significant. Third, game theoretical analysis of movement development provides opportunities to apply many other research methods—such as a comparative static approach and elasticity as a general term for a rate of change—to the study of social movements. Finally, the concept of a tradeoff between a social movement's short-term payoffs and long-term interests is new, and sheds light on the unintended outcomes of social movements.

6 Information gap and bloody confrontation

The final game

The social movement progresses to the last round of the repeated games, at which point the newly created social system appears to develop a fourth property—namely, an information gap between the movement and the state, with bloody confrontation as the Nash equilibrium. What is this information gap between the movement and the state?

In 1989 China, for example, when the most senior leaders finally decided to end the movement by military crackdown, they knew that suppressing the movement would almost certainly be followed by condemnation and serious economic sanctions from the international community. By way of damage control, they wanted to avoid violence and minimize casualties. Therefore, government officials repeatedly signaled to the activists that they intended to use force before they ordered troops to open fire.

In the afternoon of 3 June 1989, several hours before the military crackdown, the headquarters of the martial law troops and the municipal government of Beijing had issued some public notices and this announcement: "The situation in Beijing at present is very serious. More serious riots can occur at any time. From now on, please do not come onto the street and do not go to Tiananmen Square" (*China Daily*, 4 June 1989). At 6:30 p.m., a few hours before the crackdown began, all government-controlled radio and television began broadcasting this announcement repeatedly throughout the city and the evening, warning people not to go out on the streets or "they should bear responsibility for their own fates" (*China Daily*, 4 June 1989).

The Chinese government sent its message about the pending military action loudly and clearly. The news that the government was planning to use force spread quickly and widely among all movement participants in Beijing. How did the students and their supporters take the news? Did they prepare themselves for the military crackdown psychologically and practically? All evidence shows that they never took the threat seriously and even excluded a military crackdown as a possible outcome.

When the crackdown began at Muxudi (an intersection on West Changan Boulevard, a few miles away from Tiananmen Square) at about 10:30 that evening, the opening ceremony of a new university—the Democracy University, which was established by the students themselves—was in progress at the square. The large crowd was very enthusiastic. More than 100,000 people had gathered in the square, and most of them sensed that the government might be planning to use new methods to enforce martial law, while they simultaneously assumed that the People's Liberation Army would never turn its guns on the people. In other words, although the students and their supporters had clearly and repeatedly heard the government's warning, they chose to ignore it.

As a concept, "information gap" describes the difference between messages sent by a state and the beliefs of movement participants—in this case, the gap between the fact that a military crackdown was about to begin and the movement's belief that massacre was unthinkable and impossible. The information gap is what makes bloody confrontation the Nash equilibrium in the final game, and it is what generates unintended outcomes of social movements.

Why is there an information gap between the movement and the state? If Chinese government officials had really wanted to warn the students that a military operation was about to begin, why couldn't they have undertaken strenuous efforts to close the information gap? Why were government officials, who controlled all news media, incapable of persuading the students that the state intended to use force? Was the students' belief that a massacre would never occur a false expectation, or an accurate one? What was the foundation for the formation of the students' perceptions?

Almost all of these questions have been disregarded by previous studies of the 1989 Tiananmen tragedy, although numerous works on the subject have been published. This chapter provides answers that help to explain what really took place behind the military crackdown in the last stage of the movement's development.

State's threats: interdependent or independent

In the first section of Chapter 5, I discussed how the students and their supporters drew inferences and formed beliefs about the state's preferences by observing the state's words and deeds, and how they modified their beliefs in the repeated game. The movement's development is a process in which participants continually revise their beliefs, and it is also a process in which government forms its credibility—i.e., its capability to be believed or trusted.

The Chinese government threatened the movement three times, and backed down after students and their supporters resisted the first and second threats. Movement participants sagaciously observed that the state's credibility was low, and their perception was supported by normal daily experience. As Christopher J. Fettweis describes:

> Individuals certainly develop reputations in their daily lives that influence the way that others treat them. Parents understand that they must carry through on their threats and promises if they want their children to take their future instructions seriously, and we all have friends whose repeated failures to deliver on past promises make us skeptical of their future assurances.
>
> (2007:618)

It has been demonstrated that this credibility imperative has an enormous effect on foreign policy and that many parallels exist between the impact of a state's credibility on its international relationships and its impact on social movement development. According to Fettweis:

> Experienced practitioners of foreign policy take for granted the notion that actions taken today can affect (and perhaps prevent) the crises of tomorrow. The messages sent by foreign policy actions can sometimes seem to be more important than the actions themselves, since other states—including current and potential enemies—are watching every move, making judgments about the credibility of U.S. threats and promises.
>
> (2007:607)

This credibility is, "in Kissinger's words, 'the coin with which we conduct our foreign policy,' an intangible asset without which a state cannot influence the actions of others" (Fettweis, 2007:609).

How does change in the state's level of credibility generate the information gap? Did the Chinese government base its strategies on the imperative to remain credible to the 1989 movement? Before answering these questions, we need to be aware that serious debate exists between scholars and practitioners over the importance of a state's credibility. The lessons we

Table 6.1 A debate over state credibility

Comparison between:	*The credibility imperative*	*A competing conventional wisdom*
The importance of state's credibility	State's credibility is an intangible asset	State's credibility is an illusion–and an exceptionally dangerous illusion at that
	The credibility of a state serves as a guide for others trying to anticipate future actions	States cannot control their level of credibility
The fundamental assumption	An essential interdependence exists between foreign policy actions across time and space	Foreign policy actions are far more independent than is commonly believed
The importance of messages sent by a state's action	The actions of a state send messages to others about its probable behavior in the future	Sending messages for others' consideration in future crises is all but futile
Effects of the state's credibility on policy	High credibility helps a state achieve its goals	Crisis behavior appears to be entirely independent; credibility is, therefore, all but irrelevant

draw from both sides of this debate can deepen our understanding of the last stage of the 1989 movement. Fettweis describes the debate as follows:

> This "credibility imperative", in historian Robert McMahon's words, has occupied a central position in every major foreign policy debate in the last 50 years. However, many scholars remain, for the most part, unconvinced. Since the war in Vietnam, a competing conventional wisdom has been building momentum in scholarly circles, one that considers the obsession with credibility to be an illusory waste of time at best and a producer of profoundly destructive misguidance at worst. Few issues currently separate the policy community from scholars in a more profound and important way. Resolving this curious disconnect between the beliefs of scholars and those of practitioners could hardly be more urgent.
>
> (2007:607)

What is the competing conventional wisdom? Can we really explain the information gap between the state and movement by the competing conventional wisdom instead of the credibility imperative?

In Table 6.1, I summarize the main points of both sides in the debate over state credibility, based on Fettweis's 2007 paper, "Credibility and the War on Terror."

Fettweis strongly believes in the competing conventional wisdom, and argues: "When the credibility imperative drives policy, due to the inherently uncontrollable nature of this most intangible of assets, states march toward folly" (Fettweis, 2007:633). I challenge this competing conventional wisdom, however, and will argue that the state's credibility has a significant impact on the dynamics of social movements. More specifically, I demonstrate that states march toward folly (i.e., that they cannot control movement development) when the credibility imperative does not drive their strategy.

In Chapter 5, I described movement development as repeated games involving three rounds. The government threatened the movement in each round. Are the state's three threats interdependent or independent? The evidence strongly supports their interdependence, since the state's choices in the first round (to threaten first, show force, and back down if the students resist the threat) affect its feasible choices in the last round (to threaten first, show force, and then use force if students resist the threat). In Chapter 5, I demonstrated how the state's choice in the last round of the repeated game was affected by its choice in the first round. In other words, there is no evidence in social movement development to support the competing conventional wisdom's fundamental assumption that the state's threats are independent.

State's interests: the tangible interests of the present or the intangible interests of the future

Whenever state threats are interdependent across time, government officials have to take into account the impact of their current action—i.e., the impact of the tangible interests of the present—on the intangible interests of the future. In the case of the 1989 Chinese student movement, the first round of the repeated game is viewed as the present and the second and third rounds represent the future. If the state's payoff in the first round is the tangible interests of the present (i.e., that the Chinese government maintains legitimacy by backing down) then the state's credibility lies in the intangible interests of the future, because it has significant impacts on the future actions of the students and their supporters.

Both the tangible interests of the present and the intangible interests of the future are important. The state must therefore balance these two factors in the decision-making process. How did the Chinese government balance them when officials chose their strategies in relation to the movement's development?

In the first round of the repeated game, thousands of armed police withdrew from their positions after the students resisted the state's first threat. In the second round of the repeated game, the troops were called off when 1,000,000 people protested against the implementation of martial law. It is obvious that the Chinese government failed to rise to the movement's challenge, which means that government officials ranked the tangible interests of the present more highly than the intangible interests of the future, and therefore risked earning a reputation for weakness.

Fettweis, however, does not think that the two factors should be balanced: "As unsettling as it may be, the future is largely outside our control: the tangible interests of the present, therefore, must outweigh the intangible interests of the future" (2007:633). "Keeping an eye on the future, although natural and comforting, often creates profound myopia in the present" (2007:608). Is it rational to rank the tangible interests of the present above the intangible interests of the future? Does keeping an eye on the future really create profound myopia in the present?

Let us look at empirical evidence to decide whether states should balance the two factors in their decision-making process. Let us consider the consequences of ignoring the intangible interests of the future. In the 1989 Chinese student movement, when government officials were obsessed with the tangible interests of the present in the first and the second rounds of the repeated game, two serious consequences resulted.

First, the government's credibility had significantly declined, and the information gap emerged as a result. How to measure credibility? According to Fettweis, credibility is immeasurable: "All psychological variables are difficult to measure and inherently unquantifiable: credibility is no different" (2007:610). I challenge Fettweis, and will show that credibility is measurable and that a decline in state credibility generates the information gap in relation to social movement development. I have demonstrated the following equation as equation (7) in Chapter 5:

$$\frac{P(H|D)}{P(-H|D)} = \frac{P(H)}{P(-H)} \times \frac{P(D|H)}{P(D|-H)}$$

In this equation, as I mentioned in Chapter 5, P denotes participants' beliefs about the probability that H or –H is true. H represents the hypothesis that a state will not use force after students resist its threat, –H is the hypothesis that a state will use force after students resist its threat, and D denotes data—namely, the new information observed by students that the state did not use force.

On the left side of this equation, $\frac{P(H|D)}{P(-H|D)}$ are the posterior odds that the hypothesis, H, is true, conditional upon having observed the data, D. As the posterior odds increase, the probability increases that H is true. In other words, the greater the posterior odds, the more likely it is that students and their supporters will believe that the state is not going to use force. The smaller the posterior odds, the more likely it becomes that students will believe that the state will use force.

On the right side of this equation, $\frac{P(H)}{P(-H)}$ are the prior odds that the hypothesis, H, is true. $\frac{P(D|H)}{P(D|-H)}$ represent the likelihood ratio that the data, D, would be observed if H were true rather than false. The greater the likelihood ratio, the more likely it is that the hypothesis, H, is true.

It is obvious that the posterior odds depend upon two variables: the prior odds and the likelihood ratio. In repeated games, participants' posterior beliefs in the first round become their prior beliefs in the second round, and so on. In other words, movement development is a learning process in which participants continually revise their prior beliefs and form their posterior beliefs.

In Chapter 5, I discussed how the Chinese students and their supporters had revised their prior beliefs in light of the new information (see Table 5.2, in Chapter 5). In order to measure a state's credibility, it is necessary to emphasize three conclusions that I have drawn from Table 5.2:

1 in the first round of the repeated game, participants believe in a 50/50 chance that the hypothesis is true—namely, that the state will not use force;
2 in the second round, they believe that the chance is greater than 50 percent that the hypothesis is true;
3 while participants' belief that the hypothesis is true gradually increases, their belief that the state will use force significantly decreases.

It is very important to recognize that movement development is not only a learning process for participants, but that it is also an assessing

Table 6.2 The measurement of state credibility

Repeated game	*State's credibility (C)*	*Prior odds (Prior)*	*Likelihood ratio (LR)*	*Posterior odds (Posterior)*
The first round	$C_1 = 50\%$	$Prior_1 = 50/50 = 1$	$LR_1 > 1$	$Posterior_1 > Prior_1$
The second round	$C_2 < C_1$ $C_2 < 50\%$	$Prior_2 > Prior_1$ $Prior_2 > 1$	$LR_2 > 1$	$Posterior_2 > Prior_2$
The third round	$C_3 < C_2 < 50\%$	$Prior_3 > Prior_2 > 1$	$LR_3 = 0$	$Posterior_3 = 0$

process in which participants continually measure the state's credibility (see Table 6.2).

Since Table 6.2 is based on Table 5.2, I will omit explanations of Table 5.2 in elaborating upon Table 6.2. In the first round of the repeated game, the Chinese government sent a signal that the student protests would no longer be tolerated. Use of the word "turmoil" in the official editorial was heavily loaded and was used by the government to justify any crackdown it might choose to make. Students were uncertain about whether the government would maintain credible commitments, and they believed that there was a 50 percent chance that the government could be trusted.

In the second round, government officials asserted their strong intention to use force. Declaration of martial law and the deployment of thousands of troops accompanied by tanks very clearly demonstrated the government's preferences, but students were still uncertain whether the government would pursue its goal with sufficient resolve. Unlike before, when the students and their supporters in this round measured the state's credibility, they took into account the new information: $LR_1 > 1$, which meant that the government had backed down and lost its credibility in the first round. For movement participants, the chance that the government could be trusted in the second round was therefore less than 50 percent.

In the third round, after top leaders had finally decided to end the movement by military crackdown, government officials repeatedly signaled activists about their intention to use force. At that time, however, the state's credibility had reached its lowest point during the movement's development as a result of $LR_2 > 1$: the government had once again backed down and lost even more credibility in the second round. The reason for $LR_3 = 0$ is that the government used force in the third round. See the following:

$$\because \; P(D|H) = 0$$

$$\therefore \; \frac{P(D|H)}{P(D|-H)} = 0$$

$$\therefore \; \frac{P(H)}{P(-H)} \times \frac{P(D|H)}{P(D|-H)} = 0$$

Table 6.2 clearly shows that the state's credibility significantly declined during the movement's development: $C_1 > C_2 > C_3$. Threats made by a government without credibility may not be believed. Therefore, the information gap emerges, which is the first tangible consequence of ignoring the intangible interests of the future in the state's decision-making process.

Second, there is evidence of a domino effect once the state's credibility has significantly declined. In the last stage of the 1989 Chinese student movement, for example, the government's credibility had reached its lowest point and state authorities had lost all means of showing their preference to use force when they wanted to send a clear message to the movement, even though they controlled all mass media. In other words, the media became ineffective at that time in China. No matter how hard government officials tried, they could no longer close the information gap. The situation continually deteriorated. Although the Chinese government had controlled its use of strong military force until then, its lack of credibility paralyzed officials when it came to controlling the dynamics of the social movement.

On 3 June 1989, when troops armed with tanks, assault rifles, and machine guns had arrived at Muxudi, a few miles away from Tiananmen, more than 100,000 students and their supporters were still gathered in the square. The unintended confrontation became inevitable, which was the worst nightmare for government officials, who understood that bloodshed was sure to inflame the international community and result in the imposition of economic sanctions and the condemnation of China.

In short, the state put the tangible interests of the present above the intangible interests of the future in its decision-making process, and serious consequences resulted. The government's credibility, in terms of the people's readiness to believe or trust it, significantly declined. When credibility is low, a state's threats may be ignored, and an information gap between the state and movement emerges. Like dominoes falling, a state backed by strong military force earns a reputation for weakness if it does not follow through on its threats; and state authorities lose all means to communicate with the movement—and even lose control of the movement's dynamics.

It therefore becomes clear that the credibility imperative has an enormous effect on the dynamics of a social movement, and, indeed, that movement development is impossible to understand without it. Although there is no clear way for any state to control the perception of movement participants, sending messages for participants' consideration in future development is extremely important. In the following section, I will therefore show how the credibility of a state serves as one of two important guides for movement participants who try to anticipate future state actions.

The foundation for the formation of the participants' beliefs: past experience or a correct understanding of the current situation

Almost all the people who participated in the 1989 Chinese student movement expected that the government would back down in the third round of the repeated game. Although the movement's prior experience in the first and the second rounds of the repeated game provided a powerful intuitive logic for participants to ignore the government's warning that a military crackdown was about to be adopted, history has shown their belief in the impossibility of a massacre to be a false expectation. Why was their expectation false, when they formed their beliefs based on past experience?

Daryl Press studies the impact of state credibility on its opponent's behavior. He argues:

> Children use past actions when they evaluate their parents' credibility to punish them, and perhaps we all use past actions to assess whether a friend will show up at the movies. But there is no logical basis to generalize from these mundane situations to the most critical decisions made by national leaders during crises. In fact it would be odd—even irrational—if people relied on the same mental shortcuts that they use to make unimportant split-second decisions of daily life when they confront the most important decisions of their lives—decisions on which their country's survival depends.
>
> (2006:12)

Fettweis believes that "Press argues that national capabilities and interests—not past behavior—provide the foundation for the formation of perceptions" (Fettweis, 2007:619).

According to Press, there are two kinds of decisions—important ones and unimportant ones—and important decisions cannot be made based on past experience, while it is common to base the unimportant ones on past actions. Press's argument generates a logical inconsistency, however: why do people follow a different logic when they make important and unimportant decisions? Why does the same decision-making process based on past actions generate rational decisions when these decisions are unimportant—and irrational decisions when the decisions are important?

I challenge Press, and I will argue that two factors provide the foundation for the formation of perceptions: past experience and current situation. Both of them should be taken into account in the decision-making process, whether or not the decision is important. In other words, it was not wrong for the students and their supporters to form their beliefs in the third round of the repeated game based on their past experience that the Chinese government's credibility had significantly declined during the movement's development. But it was wrong for them to ignore the current situation.

What is the current situation? As Press points out, it is the opponent's capabilities and interests. In the case of the 1989 Chinese student movement, for example, when students and their supporters assessed whether the government would use force in the third round, they should have asked themselves at least three questions:

1 What had the government done in the first and the second rounds?
2 Was the government capable of adopting a policy to implement a military crackdown in the third round?
3 Was it in the government's interest to end the movement by military crackdown after two rounds of the repeated game? The first question relates to past experience, and the second and third relate to the situation at hand.

Unfortunately, the students and their supporters focused all their attention on the first question. The common reaction to the government's warning was that the government might be planning to use some new methods to enforce martial law, but that none of them would work because nothing dire had happened since martial law was implemented.

Their obsession with past experience blinded the students and their supporters to the fact that reality had changed after the first two rounds. They overlooked the government's messages and the many signs of government preparation for the military crackdown. For instance, although they had witnessed military equipment being transferred into the city even before the government broadcast the final announcement, few of them thought that it was a harbinger of violence. They also lacked knowledge about the dynamics of social movements, such as the systematic connection between the increase in their payoffs and an escalation in the probability that the state would use force. Their limited knowledge rendered them incapable of analyzing how changes in the government's interests could lead to a military crackdown.

In sum, the students' belief that massacre would never occur was a false expectation in the third round of the repeated game. It is not wrong to interpret the validity of state threats on the basis of the government's credibility, but the formation of an accurate expectation depends not only upon past experience, but also on the correct understanding of a current situation. Unfortunately, sociology's failure to provide knowledge about unintended outcomes of social movements makes it virtually impossible for movement participants to correctly understand the dynamics of movement development.

The following proposition about the strategy chosen by the participants in the final game is drawn from this chapter:

> *Proposition 4: in the final game, the state clearly signals to the movement participants its intention to use force, but participants choose to ignore the government's warning due to the existence of an information gap between the movement and the state. The participants continually resist the state's threat, as a result of which bloody confrontation becomes an unintended outcome of social movement.*

Conclusion

The important finding in this chapter is that an information gap between the movement and the state emerges in the last stage of movement development. It is this information gap that makes bloody confrontation the Nash equilibrium in the final game. It is the information gap that generates unintended outcomes of social movements.

There are two causes of this gap. The first is that the state fails to choose strategies based on the credibility imperative. As I demonstrated

Table 6.3 Comparison between game theoretical approach and other existing approaches

Comparison between:	*Game theoretical approach*	*Other existing theoretical approaches*
Is the state's credibility measurable?	Yes, it can be measured	No, it is immeasurable
If a state controls all mass media, does this mean that it can effectively communicate with a movement?	No. State behavior in the first round of the repeated game influences its behavior in the last round. When the state's credibility is low, it loses all means to communicate with a movement	Yes. Study of the domino effect of decline in state credibility remains to be undertaken
If a state controls its strong military force, does this mean that it can control movement development?	No. State's threats are interdependent, and lack of credibility can paralyze the state when it comes to controlling the dynamics of social movements	Yes. Study of the domino effect of decline in state credibility remains to be undertaken
Policy implication	For the state: • strategies chosen in the first round of the repeated game can definitely affect the crises of the last round • messages sent by state strategies can sometimes seem more important than the actions themselves For the movement: • the formation of an accurate expectation depends on past experience and the correct understanding of a current situation	

in Chapter 3, states usually issue threats against participants after the emergence of a social movement; however, such threats cannot hinder the social movement. If the state accepts status quo ante—insofar as the state does not follow up its promises—its credibility becomes low, and this decrease in credibility has a domino effect on movement development. With each repetition of this cycle, the situation deteriorates and the state's credibility continually declines until an information gap between the movement and the state emerges.

The second cause is that movement participants fail to correctly understand their current situation. They base their perception and beliefs about the state exclusively on their past experience that the state's credibility has significantly declined during the movement's development. But they choose to ignore the state's capabilities and interests.

Can this information gap be closed in the last stage of movement development? It seems that both the state and the movement participants are incapable of closing it. State authorities have lost all means to communicate with the movement, and participants lack the knowledge and capability to correctly understand the dynamics of movement development. Bloody confrontation in the final game is therefore the price that is paid for closing the information gap.

Finally, it is necessary to compare the game theoretical approach and the existing approach to the study of the final game (see Table 6.3 on the previous page).

This comparison shows that the game theoretical approach has significant advantages for studying the enormous effect of the credibility imperative on social movement development. First, while other scholars argue that the state's credibility is immeasurable, I demonstrated how movement participants measure the state's credibility round by round in the repeated games. Second, some studies demonstrate that crisis behavior appears to be entirely independent, but the game theoretical analysis of movement development shows that the movement's crisis behavior in the final game is strongly influenced, not only by the state's strategies in the first and second rounds, but also by a domino effect produced by decline in the state's credibility. Finally, the concept of an information gap between the movement and the state is a new idea that sheds light on the unintended outcomes of social movements.

Appendix

A mixed strategy Nash equilibrium in Figure 5.1

Let the state play "Sub-optimal Strategy" with the probability m, and play "Optimal Strategy" with the probability 1-m.

When movement participants choose "Resist," their payoff is the following:

$$mX3 + (1-m)X4 = mX3 + X4 - mX4$$

When movement participants choose "Back Down," their payoff is the following:

$$mX2 + (1-m)X2 = X2$$

Let movement participants be equally inclined to choose "Resist" and "Back Down," then,

$$mX3 + X4 - mX4 = X2$$

$$m(X3 - X4) = (X2 - X4)$$

$$m = \frac{X2 - X4}{X3 - X4}$$

$$(1-m) = \frac{X3 - X2}{X3 - X4}$$

Let movement participants play "Resist" with the probability n, and play "Back Down" with the probability 1–n.

When the state chooses "Sub-optimal Strategy," its payoff is the following:

$$nY3 + (1 - n)Y2 = nY3 + Y2 - nY2$$

When the state chooses "Optimal Strategy," its payoff is the following:

$$nY4 + (1 - n)Y2' = nY4 + Y2' - nY2'$$

Let the state be equally inclined to choose "Sub-optimal Strategy" and "Optimal Strategy," then,

$$nY3 + Y2 - nY2 = nY4 + Y2' - nY2'$$

$$nY3 - nY2 - nY4 + nY2' = Y2' - Y2$$

$$n(Y3 - Y2 - Y4 + Y2') = (Y2' - Y2)$$

$$n = \frac{Y2' - Y2}{Y3 - Y2 - Y4 + Y2'} = \frac{Y2 - Y2'}{Y2 - Y2' - Y3 + Y4}$$

$$(1 - n) = \frac{Y4 - Y3}{Y2 - Y2' - Y3 + Y4}$$

Conclusion

In the mixed strategy Nash equilibrium, the state plays "Sub-optimal Strategy" with the probability $\frac{X2-X4}{X3-X4}$ and plays "Optimal Strategy" with the probability $\frac{X3-X2}{X3-X4}$.

In the mixed strategy Nash equilibrium, movement participants play "Resist" with the probability $\frac{Y2-Y2'}{Y2-Y2'-Y3+Y4}$ and play "Back Down" with the probability $\frac{Y4-Y3}{Y2-Y2'-Y3+Y4}$.

Bibliography

Altman, Roger C. 2009. "The Great Crash, 2008." *Foreign Affairs* 88:2–14.

Amenda, Edwin, and Yvonne Zylan. 1991. "It Happened Here: Political Opportunity, the New Institutionalism, and the Townsend Movement." *American Sociological Review* 56:250–65.

Amenda, Edwin, Bruce G. Carruthers, and Yvonne Zylan. 1992. "A Hero for the Aged? The Townsend Movement, the Political Mediation Model, and U.S. Old-Age Policy, 1934–1950." *American Journal of Sociology* 98:308–39.

Amenda, Edwin, Kathleen Dunleavy, and Mary Bernstein. 1994. "Stolen Thunder? Huey Long's 'Share Our Wealth', Political Mediation, and the Second New Deal." *American Sociological Review* 59:678–702.

Anderson, Margaret L. and Howard F. Taylor. 2007. *Sociology: The Essentials*. Boston, MA, Wadsworth Publishing.

Azjen, I. and M. Fishbein. 1975. "A Bayesian Analysis of the Attribution Process." *Psychological Bulletin* 82:261–77.

Barkey, Karen. 1991. "Rebellions Alliances: The State and Peasant Unrest in Early Seventeenth-Century France and The Ottoman Empire." *American Sociological Review* 56: 699–715.

Blaug, Mark. 1992. *The Methodology of Economics or How Economists Explain*. Cambridge: Cambridge University Press.

Boswell, Terry, and William J. Dixon. 1993. "Marx's Theory of Rebellion: A Cross-National Analysis of Class Exploitation, Economic Development, and Violence Revolt." *American Sociological Review* 58:681–702.

Brinton, Crane. 1968. *The Anatomy of Revolution*. New York: W.W. Norton.

Brook, Timothy. 1998. *Quelling the People: The Military Suppression of the Beijing Democracy Movement*. Stanford, CA: Stanford University Press.

Burstein, Paul. 1991. "Legal Mobilization as a Social Movement Tactic: The Struggle for Equal Employment Opportunity." *American Journal of Sociology* 96:1201–25.

Chiang, Alpha C. 1984. *Fundamental Methods of Mathematical Economics*. New York: McGraw-Hill, Inc.

China Daily. April 1989 – June 1989. Beijing, China.

China Statistical Yearbook. 2009. Beijing: China Statistics Press.

Chong, Dennis. 1991. *Collective Action and the Civil Rights Movement*. Chicago: University of Chicago Press.

Church, George J. 2005. "China." *Time.com*: http://www.time.com/time/printout/0, 8816,1074879,00.html [accessed June 23, 2009].

Coleman, James S. 1990. *Foundations of Social Theory*, Cambridge, MA: Belknap Press of Harvard University Press.

——. 1994. "A Vision for Sociology." *Society* 32(1):29–34.

Conell, Caro, and Kim Voss. 1990. "Formal Organization and the Fate of Social Movements: Craft Association and Class Alliance in the Knights of Labor." *American Sociological Review* 55:255–69.

Davis, James C. 1971, *When Men Revolt and Why*, New York: Free Press.

Deng, Fang. 1997. "Information Gap and Unintended Outcomes: The Chinese Student Movement." *American Journal of Sociology* 102:1085–112.

Ding, X. L. 1994. *The Decline of Communism in China*. Cambridge, MA: Cambridge University Press.

Edward, L. P. 1927. *The Natural History of Revolution*. Chicago: University of Chicago Press.

Fearon, James Dana. 1992. *Threats to Use Force: Costly Signals and Bargaining in International Crises*. Ann Arbor, MI: U.M.I. Dissertation Services, A Bell and Howell Company.

Fettweis, Christopher J. 2007. "Credibility and the War on Terror." *Political Science Quarterly* 122:607–33.

Frey, R. Scott, Thomas Dietz, and Linda Kalof. 1992. "Characteristics of Successful American Protest Groups: Another Look at Gamson's Strategy of Social Protest." *American Journal of Sociology* 98:368–87.

Gamson, William, 1975. *The Strategy of Social Protest*. Homeward, IL: Dorsey.

Gerhards, Jurgen, and Dieter Rucht. 1992. "Mesomobilization: Organizing and Framing in Two Protest Campaigns in West Germany." *American Journal of Sociology* 98:555–95.

Gibbons, Robert, 1992. *Game Theory for Applied Economics*. Princeton, NJ: Princeton University Press.

Goldstone, Jack A. 1996. "State Breakdown in the English Revolution: A New Synthesis." *American Journal of Sociology* 92:257–322.

Gould, Roger V. 1991. "Multiple Networks and Mobilization in the Paris Commune, 1871." *American Sociological Review* 56:716–29.

——. 1993. "Trade Cohesion, Class Unity, and Urban Insurrection: Artisanal Activism in the Paris Commune." *American Journal of Sociology* 98:721–54.

——. 1993. "Collective Action and Network Structure." *American Sociological Review* 58:182–96.

Grant II, Don Sherman, and Michael Wallace. 1991. "Why Do Strikes Turn Violent?" *American Journal of Sociology* 96:1117–50.

Gurr, Ted. 1970. *Why Men Rebel*. Princeton, NJ: Princeton University Press.

Heckathorn, Douglas D. 1988. "Collective Sanctions and the Creation of Prisoner's Dilemma Norms." *American Journal of Sociology* 94:535–62.

Hedstrom, Peter. 1994. "Contagious Collectives: On the Spatial Diffusion of Swedish Trade Unions, 1890–1940." *American Journal of Sociology* 99:1157–79.

Hirsch, Eric L. 1990. "Sacrifice for the Cause: Group Processes Recruitment, and Commitment in a Student Social Movement." *American Sociological Review* 55:243–54.

Hirshleifer, Jack, and Amihai Glazer. 1992. *Price Theory and Applications*. Englewood Cliffs, NJ, Prentice-Hall, Inc.

Hirshleifer, Jack, and John G. Riley. 1992. *The Analytic of Uncertainty and Information*. Cambridge, MA: Cambridge University Press.

Jenkins, J. Craig, and Craig M. Eckert. 1986. "Channeling Black Insurgency: Elite Patronage and Professional Social Movement Organization in the Development of the Black Movement." *American Sociological Review* 51:812–29.

Karstedt-Henke, Sabine. 1980. "Theories for the Explanation of Terrorist Movements." In *The Politics of Internal Security*, edited by E. Blankenberg. Frankfurt: Suhrkamp, pp. 178–96.

Kesselman, Mark, Joel Krieger, and William A. Joseph. 2003. "Introducing Comparative Politics." In *Introduction to Comparative Politics*, edited by Mark Kesselman, Joel Krieger, and William A. Joseph. Boston, MA: Wadsworth Publishing, pp. 3–47.

Kitschelt, Herbert P. 1986. "Political Opportunity Structures and Political Protest: Anti-Nuclear Movement in Four Democracies." *British Journal of Political Science* 16: 57–85.

Klandermans, Bert. 1994. "Mobilization and Participation: Social Psychological Expansions of Resource Mobilization Theory." *American Sociological Review* 49:583–600.

Klandermans, Bert, and Dirk Oegema. 1987. "Potentials, Networks, Motivations, and Barriers: Steps Towards Participation in Social Movements." *American Sociological Review* 52:519–31.

Knoke, David. 1988. "Incentives in Collective Action Organizations." *American Sociological Review* 53:311–29.

Koopmans, Ruud. 1993. "The Dynamics of Protest Waves: West Germany, 1965 to 1989." *American Sociological Review* 58:637–56.

Kreps, David M. 1990a. *A Course in Microeconomic Theory*. Princeton, NJ: Princeton University Press.

——. 1990b. *Game Theory and Economic Modeling*. New York: Oxford University Press.

Kreps, D. and R. Wilson. 1982. "Sequential Equilibrium." *Econometrica* 50:863–94.

Lewis, Jerry M. and Thomas R. Hensley. 2009. "The May 4 Shootings at Kent State University: The Search for Historical Accuracy." http://dept.kent.edu/sociology/lewis/lewihen.htm [accessed April 15, 2010].

McAdam, Dong. 1982. *Political Process and the Development of Black Insurgency, 1930–1970*. Chicago: University of Chicago Press.

——. 1983. "Tactical Innovation and the Pace of Insurgency" *American Sociological Review* 48:735–54.

——. 1986. "Recruitment to High-Risk Activism: The Case of Freedom Summer." *American Journal of Sociology* 92:64–90.

——. 1988. *Freedom Summer: The Idealists Revisited*. New York: Oxford University Press.

——. 1989. "The Biographical Consequences of Activism." *American Sociological Review* 54:744–60.

McAdam, Dong, and Ronnelle Paulsen. 1993. "Specifying the relationship between Social Ties and Activism." *American Sociological Review* 99:640–67.

McCarthy, John D. and Mayer N. Zald. 1977. "Resource Mobilization and Social Movements: A Partial Theory." *American Journal of Sociology* 82:1212–41.

——. 1973. *The Trends of Social Movement in America Professionalization and Resource Mobilization*. Morristown, NJ: General Learning Press.

Macy, Michael W. 1990. "Learning Theory and the Logic of Critical Mass." *American Sociological Review* 55:809–26.

——. 1991. "Chains of Cooperation: Threshold Effects in Collective Action." *American Sociological Review* 56:730–47.

Markoff, John. 1985. "The Social Geography of Rural Revolt at the Beginning of the French Revolution." *American Sociological Review* 50:761–81.

——. 1986. "Literacy and Revolt: Some Empirical Notes on 1789 in France." *American Journal of Sociology* 92:323–49.

Marwell Gerald, Pamela E. Oliver, and Ralph Prahl. 1988. "Social Networks and Collective Action: A Theory of the Critical Mass. III." *American Journal of Sociology* 94:502–34.

Moaddel, Mansoor. 1992. "Ideology as Episodic Discourse: The Case of the Iranian Revolution." *American Sociological Review* 57:353–79.

Moore, Will H. 1995. "Rational Rebels: Overcoming the Free-Rider Problem." *Political Research Quarterly* 48:417–54.

Morris, Aldon D. 1981. "Black Southern Student Sit-In Movement: An Analysis of Internal Organization." *American Sociological Review* 46:755–67.

——. 1983. "Birmingham Confrontation Reconsidered: An Analysis of the Dynamics and Tactics of Mobilization." *American Sociological Review* 58:621–36.

——. 1984. *Origin of the Civil Rights Movement: Black Communities Organizing for Change*. New York: Free Press.

Morris, Aldon D., and Cedric Herring. 1987. "Theory and Research in Social Movements: A Critical Review." In *Political Behavior Annual*, edited by Samuel Long. Boulder, CO: Westview Press, pp. 103–29.

Morrow, James D. 1994. *Game Theory for Political Scientists*. Princeton, NJ: Princeton University Press.

Mu, Yi, Mark V. Thompson, and China Dissident Collection. 1990. *Crisis at Tiananmen: Reform and Reality in Modern China*. San Francisco, CA: China Books and Periodicals, Inc.

Mueller, Edward N. 1985. "Income Inequality, Regime Repressiveness, and Political Violence." *American Sociological Review* 50:47–61.

Myerson, Roger B. 1991. *Game Theory: Analysis of Conflict*. Cambridge, MA: Harvard University Press.

Oberschall, Anthony, 1973. *Social Conflict and Social Movements*. Englewood Cliffs, NJ: Prentice-Hall.

Oegema, Dirk, and Bert Klandermans. 1994. "Why Social Movement Sympathizers Don't Participate: Erosion and Nonconversion of Support." *American Sociological Review* 59:703–22.

Ogden, Suzanne, Kathleen Hartford, Lawrence Sullivan, and Dowid Zweig. 1992. *China's Search for Democracy: The Student and the Mass Movement of 1989*. Armonk, NY: M.E. Sharpe, Inc.

Oliver, Pamela, and Gerald Marwell. 1988. "The Paradox of Group Size in Collective Action: A Theory of the Critical Mass. II." *American Sociological Review* 53:1–8.

Oliver, Pamela, Gerald Marwell, and Ruy Teixeira. 1985. "A Theory of the Critical Mass. I. Interdependence, Group Heterogeneity, and the Production of Collective Action." *American Journal of Sociology* 91:522–56.

Olson, Mancur, 1965. *The Logic of Collective Action*. Cambridge, MA: Harvard University Press.

Opps, Karl-Dieter. 1988. "Grievances and Participation in Social Movements." *American Sociological Review* 53:853–64.

Opps, Karl-Dieter, and Christine Gern. 1993. "Dissident Groups, Personal Networks, and Spontaneous Cooperation: The East German Revolution of 1989." *American Sociological Review* 58:659–80.

Ordeshook, Peter C. 1986. *Game Theory and Political Theory*. New York: Cambridge University Press.

People's Daily. April 1989 – June 1989. Beijing, China.

Press, Daryl G. 2006. *Calculating Credibility: How Leaders Assess Military Threats*. Ithaca, NY: Cornell University Press.

Putnam, Robert D. 1988. "Diplomacy and Domestic Politics: the Logic of Two-Level Games." *International Organization* 42, 3, Summer 1988.

Quadagno, Jill. 1992. "Social Movements and State Transformation: Labor Unions and Racial Conflict in the War on Poverty." *American Sociological Review* 57:616–34.

Rosenthal, Naomi, Meryl Fingrutd, Michele Ethier, Roberta Karant, and David McDonald. 1985. "Social Movements and Network Analysis: A Case Study of Nineteenth-Century Women's Reform in New York State." *American Journal of Sociology* 90:1023–55.

Scharpf, Fritz. 1988. "A Game-Theoretical Interpretation of Inflation and Unemployment in Western Europe." *Journal of Public Policy* 7:227–57.

Shepsle, Kenneth, Arthur Denzau, and William Riker. 1985. "Farquharson and Fenno: Sophisticated Voting and Home Style." *American Political Science Review* 79:1117–34.

Shin, Gi-Wook. 1994. "The Historical Making of Collective Action: The Korean Peasant Uprisings of 1946." *American Journal of Sociology* 99:1596–624.

Snow, David A., Louis A. Zurcher, Jr., and Sheldon Ekland-Olson. 1980. "Social Networks and Social Movements: A Microstructural Approach to Differential Recruitment." *American Sociological Review* 5:787–801.

Snow, David A., E. Burke Rochford, Jr., Steven K. Worden, and Robert D. Benford. 1986. "Frame Alignment Process, Micromobilization, and Movement Participation." *American Sociological Review* 51:464–81.

Snow, David A., Sarah Anne Soule, and Hanspeter Kriesi. 2004. "Mapping the Terrain." In *The Blackwell Companion to Social Movements*, edited by David A. Snow, Sarah Anne Soule, and Hanspeter Kriesi. Oxford: Blackwell Publishing, pp. 3–17.

Spence, A.M. 1973. "Job Market Signaling." *Quarterly Journal of Economics* 87:355–74.

Staggenborg, Suzanne.1988. "The Consequences of Professionalization and Formalization in the Pro-Choice Movement." *American Sociological Review* 53:585–606.

Strauss, Robert S. 1987. "Forward." in *The Tokyo Round of Multilateral Trade Negotiations: A Case Study in Building Domestic Support for Diplomacy*. Washington, DC: Georgetown University Institute for the Study of Diplomacy.

Taylor, Michael. 1987. *The Possibility of Cooperation*. New York: Cambridge University Press.

Taylor, Michael, and Hugh Ward. 1982. "Chickens, Whales and Lumpy Goods." *Political Studies* 30:350–70.

Taylor, Verta. 1989. "Social Movement Continuity: The Women's Movement in Abeyance." *American Sociological Review* 54:761–75.

Tilly, Charles.1978. *From Mobilization to Revolution.* Reading, MA: Addison-Wesley.

——. 2004. *Social Movements, 1768–2004.* Boulder, CO: Paradigm Publishers.

Tirole, Jean. 1988. *The Theory of Industrial Organization.* Cambridge, MA: MIT Press.

Tsou, Tang. 1992. "The Tiananmen Tragedy: The State–Society Relationship, Choices, and Mechanisms in Historical Perspective." In *Contemporary Chinese Politics in Historical Perspective*, edited by Brantly Womack. Cambridge, MA: Cambridge University Press, pp. 265–327.

Turner, Ralph H. 1991. "The Use and Misuse of Rational Models in Collective Behavior and Social Psychology." *Archives Europeans de Sociologie* 32(1): 84–100.

Useem, Bert. 1985. "Disorganization and the New Mexico Prison Riot of 1980." *American Sociological Review* 50:677–88.

Walton, John, and Charles Ragin. 1990. "Global and National Sources of Political Protest: Third World Responses to the Debt Crisis." *American Sociological Review* 55:876–90.

White, Robert W. 1989. "From Peaceful Protest to Guerrilla War: Micromobilization of the Provisional Irish Republican Army." *American Journal of Sociology* 94:1277–302.

Zald, Mayer N. 1992. "Looking Backward to Look Forward." In *Frontiers in Social Movement Theory*, edited by Aldon D. Morris and Carol McClury Mueller. New Haven, CT: Yale University Press, pp. 4–28.

Zald, Mayer N. and Roberta Ash. 1966. "Social Movement Organizations: Growth, Decay and Change." *Social Forces* 44:327–41.

Zald, Mayer N. and Bert Useem. 1987. "Movement and Countermovement Interaction: Mobilization, Tactics and State Involvement." In *An Organizational Society*, edited by Mayer N. Zald and John D. MaCarthy. New Brunswick, NJ: Transaction Books, pp. 125–56.

Zhao, Dingxin. 2001. *The Power of Tiananmen: State–Society Relations and the 1989 Beijing Student Movement.* Chicago: University of Chicago Press.

Young, Earnest P. 1992. "Imagining the Ancient Regime in the Deng Era." In *Popular Protest and Political Culture in Modern China*, edited by Jeffrey N. Wasserstrom and Elizabeth J. Perry. Boulder, CO: Westview Press, pp. 179–203.

Index

Note: page numbers in **bold** refer to illustrations.